Modeling and Simulation of Mechatronic Systems using Simscape

Synthesis Lectures on Mechanical Engineering

Synthesis Lectures on Mechanical Engineering series publishes 60–150 page publications pertaining to this diverse discipline of mechanical engineering. The series presents Lectures written for an audience of researchers, industry engineers, undergraduate and graduate students.

Additional Synthesis series will be developed covering key areas within mechanical engineering.

Modeling and Simulation of Mechatronic Systems using Simscape
Shuvra Das
2020

Bifurcation Dynamics of a Damped Parametric Pendulum
Yu Guo and Albert C.J. Luo
2019

Reliability-Based Mechanical Design, Volume 1: Component under Cyclic Load and Dimension Design with Required Reliability
Xiaobin Le
2019

Reliability-Based Mechanical Design, Volume 1: Component under Static Load
Xiaobin Le
2019

Solving Practical Engineering Mechanics Problems: Advanced Kinetics
Sayavur I. Bakhtiyarov
2019

Natural Corrosion Inhibitors
Shima Ghanavati Nasab, Mehdi Javaheran Yazd, Abolfazl Semnani, Homa Kahkesh, Navid Rabiee, Mohammad Rabiee, Mojtaba Bagherzadeh
2019

Modeling and Simulation of Mechatronic Systems using Simscape
Shuvra Das

ISBN: 978-3-031-79652-4   paperback
ISBN: 978-3-031-79653-1   ebook
ISBN: 978-3-031-79654-8   hardcover

DOI 10.1007/978-3-031-79653-1

A Publication in the Springer series
SYNTHESIS LECTURES ON MECHANICAL ENGINEERING

Lecture #24
Series ISSN
Print 2573-3168   Electronic 2573-3176

Modeling and Simulation of Mechatronic Systems using Simscape

Shuvra Das
University of Detroit Mercy

SYNTHESIS LECTURES ON MECHANICAL ENGINEERING #24

ABSTRACT

Mechatronic Systems consist of components and/or sub-systems which are from different engineering domains. For example, a solenoid valve has three domains that work in a synergistic fashion: electrical, magnetic, and mechanical (translation). Over the last few decades, engineering systems have become more and more mechatronic. Automobiles are transforming from being gasoline-powered mechanical devices to electric, hybrid electric and even autonomous. This kind of evolution has been possible through the synergistic integration of technology that is derived from different disciplines. Understanding and designing mechatronic systems needs to be a vital component of today's engineering education. Typical engineering programs, however, mostly continue to train students in academic silos (otherwise known as majors) such as mechanical, electrical, or computer engineering. Some universities have started offering one or more courses on this subject and a few have even started full programs around the theme of Mechatronics. Modeling the behavior of Mechatronic systems is an important step for analysis, synthesis, and optimal design of such systems. One key training necessary for developing this expertise is to have comfort and understanding of the basic physics of different domains. A second need is a suitable software tool that implements these laws with appropriate flexibility and is easy to learn.

This short text addresses the two needs: it is written for an audience who will likely have good knowledge and comfort in one of the several domains that we will consider, but not necessarily all; the book will also serve as a guide for the students to learn how to develop mechatronic system models with Simscape (a MATLAB tool box). The book uses many examples from different engineering domains to demonstrate how to develop mechatronic system models and what type of information can be obtained from the analyses.

KEYWORDS

mechatronics, multi-disciplinary, modeling and simulation, MATLAB, Simscape, physical system modeling, system response, controls, time response, frequency response

Contents

Preface

Engineering systems of today are complex and multidisciplinary. Synergistic combinations of mechanical components with electronics and control software have made engineering systems a lot more efficient than they were a few years ago. New innovations in all fields of engineering, but especially in artificial intelligence and machine learning, will make engineering systems even better in future years. Traditional engineering courses place a lot of emphasis on component design and analysis and comparatively less on system design. It is necessary to change that. Engineers should be much more comfortable in systems thinking and be adept at system design. System modeling is a powerful tool to develop user expertise in system analysis, system identification, and system synthesis, three aspects of system design. In this book, attempt has been made to introduce the reader to mechatronic system modeling and simulation. We have used Simscape, a toolbox available within Matlab/SIMULINK software suite. Simscape uses the physical network approach which is very intuitive; the technique will look familiar to users who have had basic courses in engineering. After a brief introduction to the field of Mechatronics and Simscape the rest of the book is devoted to discussing model development in Mechanical, Electrical, Magnetic, and mixed domains. The last chapter discusses a number of fairly complex Mechatronic systems including one where simulation results are used to develop a rudimentary machine learning model for the engineering system.

Shuvra Das
December 2019

CHAPTER 1

Introduction to Mechatronic Systems

1.1 INTRODUCTION

The concept behind the word mechatronics is now fairly well established in the technical community. The word was first suggested by Japanese engineers in mid-1960s as a combination of electronics and mechanics to signify electro-mechanical systems. Over time the word has come to represent multidisciplinary systems. Mechatronic systems can comprise of many domains such as mechanical, electrical, hydraulic, magnetic, etc. Components from these different domains work together in a synergistic manner along with microcomputers that control or make decision about system behavior.

The most commonly used definition for a mechatronic system is:

"Synergistic Combination of precision mechanical engineering, electronic control, and intelligent software in a systems framework, used in the design of products and manufacturing processes."

It is hard to pinpoint the origin of this definition since it is found in so many different sources including the 1997 article in *Mechanical Engineering* by Steven Ashley (1997) [2]. Giorgio Rizzoni, professor at Ohio State University, have defined it as "the confluence of traditional design methods with sensors and instrumentation technology, drive and actuator technology, embedded real-time microprocessor systems, and real-time software." Other similar definitions found in literature are:

- *"the design and manufacture of products and systems possessing both a mechanical functionality and an integrated algorithmic control."* [7]

- "the interdisciplinary field of engineering dealing with the design of products whose function relies on the integration of mechanical and electronic components coordinated by a control architecture." [7]

- *"Putting Intelligence onto Physical Systems."* [7]

- "Designing intelligent machines." [7]

These similar sounding statements all convey similar concepts about the term mechatronics. The schematic in Figure 1.1 shows a variety of technical topics and how they overlap to

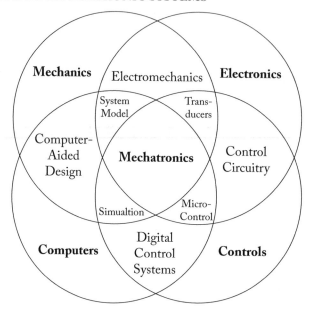

Figure 1.1: Mechatronics as an intersection of many subject areas.

define the area of mechatronics. The multidisciplinary nature of mechatronics should be obvious. What is not immediately obvious is that the concept of "synergy" is a vital part of mechatronics. Synergy implies a new way of designing these systems. In the traditional approach to electromechanical system design, the design task was undertaken as a sequential operation, i.e., the mechanical devices were designed first by mechanical engineers who then handed the design over to the electrical engineers to add on the electrical systems. The electrical engineers then passed on the design to control engineers who had to devise the control strategy. This sequential approach was also associated with very little communication among different groups and almost no effort to make each other's work better by iterating the entire system design. Synergy in mechatronics means an effective break down of the barriers between these siloed worlds. It signifies new way of doing things where engineers from different disciplines and background are involved in the product design all together and right from the beginning. This ensures a healthy dose of give and take, weighing of competing requirements, understanding each other's needs and constraints and as a result emergence of a better process and product.

Figure 1.2 shows the flow of information within a mechatronic system. At the core of the system is a mechanical system, e.g., an autonomous vehicle (Figure 1.3). In any mechatronic system, sensors sense the state of the system and the surroundings. For this particular autonomous vehicle sensors such as Lidar, camera, proximity switches, etc., are used. Information gathered by the sensors is passed onto onboard computer/microcomputer. Since sensor data is analog and computer only works with digital information analog to digital conversion is necessary prior to

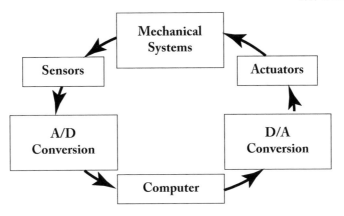

Figure 1.2: Information flow in a mechatronic system.

sending the data to the computer. Once sensor information is received by the computer it decides a course of action as per the programmed algorithm. The computer sends signals to the actuators which takes the desired action on the mechanical system. The actuators used in the case of the autonomous vehicle were DC motors attached to the wheels of the vehicle. Just like the sensor-computer interaction requires analog to digital conversion, computer-actuator interaction will require digital to analog conversion of data as well. In a way, mechatronic systems behave in a way quite similar to the biological world of living beings. For example, a human body works the same way. At the core the human body is a mechanical system. The sensors, eyes, ears, etc., gather information about the surroundings and the information is sent as signals to the brain, the computer. The brain makes decisions which are then transmitted to the muscles (the actuators); the muscles move the system in the manner desired.

Concepts of mechatronics are indispensable in today's engineering world since every product is becoming multidisciplinary in nature. If we track the evolution of a reasonably complex machine such as the automobile we will find that over the years dramatic changes have happened in automobiles with the advancement of technology. The basic functionality of an automobile, i.e., using power to drive a carriage forward was derived from the architecture of a horse-drawn carriage. Even the power used was called horse power. The horse was replaced by the internal combustion of engine. For many decades since then all developments of the automobile focused on increasing power, efficiency and output of a primarily mechanical system. With the invention of the semiconductors and IC technology the way this is achieved in an optimal manner has changed significantly. Over time and with technological advancement, less efficient sub-systems have gotten replaced by more efficient ones. Lately, this has resulted in many purely mechanical devices and sub-systems being replaced by mechatronic or electronic ones. Fuel injectors are nothing new in modern automobiles; they have replaced less efficient carburetors quite a while back. Anti-lock brakes are important safety devices and have become part of the basic pack-

Figure 1.3: Autonomous vehicle, a mechatronic system.

age for all automobiles. Similarly, "by-wire" sub-systems such as drive by-wire, brake by-wire, steer by-wire, smart suspensions, are systems that are becoming implemented as standard packages for automobiles. In all of these cases the more efficient mechatronic devices are replacing the less efficient, purely mechanical ones. It seems that we have reached the efficiency limits of purely mechanical devices. As combustion driven vehicles are being made more efficient and better controlled, electric and hybrid electric vehicles and autonomous vehicles are fast becoming commercially viable. All this has been happening because of the integration of mechanics with electronics and software in a synergistic way. Time has come perhaps for the academic world to break the same silos of separating programs as Mechanical Engineering, Electrical Engineering, etc. It is quite clear that the concept of mechatronics has evolved from being a buzzword to a very practical notion of a way of doing things driven by technological progress. Today's engineers cannot confine themselves anymore to the safe haven of their own familiar disciplines. The technological world out there will force them to venture into multidisciplinary territory and the sooner one does that the better it is.

There are many textbooks now on the topic of Mechatronics. Some of them are by authors such as Cetinkunt (2007) [5], Alciatore (2005) [1], De Silva (2005) [8], Bolton (2004) [4], Shetty and Kolk (1997) [11], Karnopp, Margolis and Rosenberg (2006) [10], Forbes T. Brown (2001) [3], and Das (2009) [6] are but a few examples.

1.2   WHAT IS A SYSTEM AND WHY MODEL SYSTEMS?

We just now discussed that at the core of the Mechatronic world is a mechanical system. We have all come across terms such as engineering systems, transmission system, transportation system, digestive system, financial system, system engineering, etc. These are phrases used in different domains, with the common theme being the concept of a "system." A system may be defined as an entity that is separable from the rest of the universe (the environment) through physical and/or conceptual boundaries. The system boundary is a logical separation between what is inside the boundary and the outside world. Although a system is separable from the surroundings it can interact with the surroundings (Karnopp et al. (2012) [10]). Systems can receive information and energy from the outside world and also send information and/or energy out (Figure 1.4). Systems may be made of interacting parts such as sub-systems and sub-systems are made of components. For example, an automobile can be considered to be an engineering system that interacts with the surroundings. It receives input from the surroundings, such as input from the driver, friction from the road, wind drag; it releases exhaust to the atmosphere, releases heat, makes noise, etc. The automobile is made of many subsystems such as drive train, transmission, brakes, etc. These subsystems are in turn made of components, such as pistons, gears, bearings, springs, latches, valves, etc. While systems are made of components (or sub-systems) a system is much more than just the sum of all its parts. Even though the parts that make up a system can be well designed and work well independently, it does not necessarily mean that the system will function well when these components are all put together. Ensuring that the system functions well after assembly is a non-trivial task and has to be done well. It is imperative that deliberate steps be taken to ensure this. For a successful final product a "systems viewpoint" is thus very important.

Systems are dynamic in nature, i.e., their behavior changes over time in response to time varying external inputs. So understanding any system's dynamic behavior is much more important than knowing its static behavior. An understanding of system behavior is a core requirement of taking a "system viewpoint." Models of systems are very useful tools for understanding dynamic behavior of systems. System models may be physical scaled models or mathematical models. Scaled physical models could be physical prototypes and may provide a hands-on understanding of system behavior. For many real-life systems building physical models may often be cost prohibitive or not possible for other reasons. At the conceptual design stage building a physical model is not possible as well. Mathematical models are much cheaper to construct but are extremely powerful if properly constructed. To build useful mathematical models one requires a good understanding of system behavior at the component level and the model builder needs to make realistic assumptions. Just as the name suggests, a model is a representation of a system but it is not necessarily the whole system. Models always involve some simplifications that are a result of simplifying assumptions made by the developer. The actual assumptions may vary from one situation to another but some of common approximations that are typically used for system modeling are:

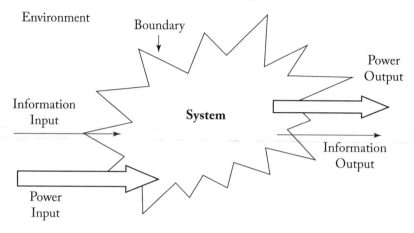

Figure 1.4: Systems.

- neglecting small effects: include the dominant effects but neglect effects that have relatively small influence;

- independent environment: the environment is not affected by what happens in the system;

- lumped characteristics: physical properties for system components are assumed to be lumped even though they are in reality distributed across the geometry;

- linear relationships: constitutive relationships are assumed to be linear over the range of operation of the system even though in reality they may not exactly be linear;

- constant parameters: parameters defining component properties are assumed to be constant; and

- neglect uncertainty and noise: any uncertainty or noise in the data are neglected.

As a result of making some of these assumptions the governing equations in the system model usually turn out to be a set of linear ordinary differential equations with constant parameters. The solutions of these ordinary differential equations are relatively easier to obtain and they describe the dynamic behavior of the system. If these simplifying assumptions are not made the equations would be a set of nonlinear partial differential equations with time and space varying parameters. This later set of equations would perhaps yield a more accurate mathematical model of the system but would not be very useful because these types of equations are much harder to solve. Without good and efficient solution techniques the model would not yield results that would be useful for engineers. The advantages gained by the simplifications far outweigh the bits of information that get lost due to these assumptions.

Mathematical System models and their solutions become powerful tools in the hands of system designers. They can be used for a variety of purposes such as the following.

- **Analysis:** For given input and known system (and state variables), what would be the output?

- **Identification:** For given input history and known output history, what would the model and its state variables be?

- **Synthesis:** For given input and a desired output, design the system (along with its state variables) so that the system performs the way desired.

Learning how to develop useful system models takes time and experience. We therefore go about the above three activities in the order that they are stated. Beginning system modelers spend a lot of time learning to "analyze" systems. Only after a good bit of experience do they venture into the world of "identification" of system. And "Synthesis" requires the maximum amount of experience in the field.

Because a model is somewhat of a simplification of the reality, there is a great deal of art in the construction of models. An overly complex and detailed model may contain parameters virtually impossible to estimate, and may bring in irrelevant details which may not be necessary. Any system designer should have a way to find models of varying complexity so as to find the simplest model capable of answering the questions about the system under study. A system could be broken into many parts depending on the level of complexity one needs. System analysis through a breakdown into its fundamental components is an art in itself and requires some expertise and experience.

In this book we will go through a systematic methodology of developing models of engineering systems so that their dynamic behavior may be studied. Unless otherwise specified we will always make the assumptions that we have discussed here. The model development will be focused mainly toward the process of analyzing system behavior. We hope that with some practice in the area of system analysis the students would be ready to start tasks in system identification and design.

1.3   MATHEMATICAL MODELING TECHNIQUES USED IN PRACTICE

Many different approaches have been used in the development of system models. One of the most common methods is deriving the state-space equations from first principles, namely from Newton's laws for mechanics, Kirchoff's voltage and current laws for electrical circuits, etc. These relationships are then numerically solved to obtain system responses. There are several graphical approaches which are quite popular among different technical communities. Linear graphs is one of them, where state-space equations are modeled as block diagrams connected by paths showing the flow of information from one block to another. Figure 1.5 shows a SIMULINK

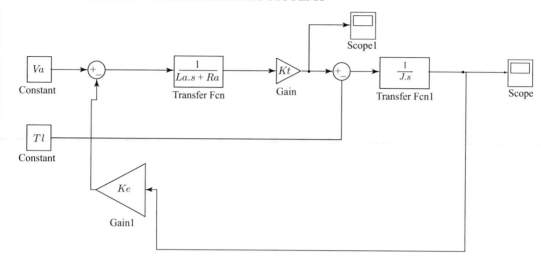

Figure 1.5: A signal flow diagram of a permanent magnet DC motor modeled in Simulink.

model of a permanent magnet DC motor built by joining different SIMULINK function blocks with proper information flow paths. This block diagram is based on signal flow, i.e., only one piece of information flows through the connections, and it is a signal rather than a physical quantity with specific meaning. The SIMULINK model is essentially a representation of the actual mathematical equations that govern the behavior of the system and in order to build such a model one has to know or derive the equations. One of the biggest challenge in building a representative system model is user-knowledge of the system. Someone who knows a system well enough including all the mathematical equations may find it easy to replicate the equations in a tool such as SIMULINK. However, someone who is encountering a system the first time may find it quite difficult to transition from a physical world to the level of abstraction necessary for a mathematical model.

An important step in all of these methods can be the derivation of the governing relationships. Within a single domain (ME, EE, etc.) deriving the governing equations may not be difficult because we may be within our specific area of expertise but when we work in a multi-domain environment it becomes somewhat more difficult for someone who is not suitably trained. The root cause of this is in how we are trained. Within each discipline of engineering education system representation and solution techniques have evolved along different paths. We are trained to think in terms of statics, dynamics, circuit analysis, electromagnetism, hydraulics, etc., to be different subject areas where different solution techniques are used for problem solving. These artificial barriers between different disciplines or subjects highlight the differences without providing a hint of the fact that the underlying similarities are much more than the perceived differences.

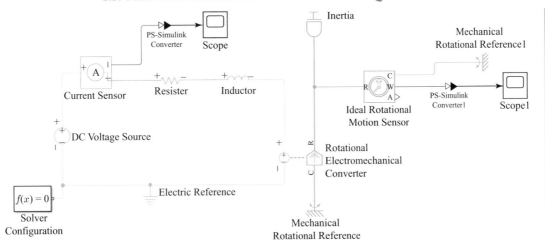

Figure 1.6: A motor model in Simscape.

There have been other tools developed which use a different approach known as physical object-based model. In this approach, components or subsystems are available as icons or blocks. Component behavior or its constitutive model is already programmed as part of the icon/block model. Thus, a capacitor block or a damper block looks similar to how they are represented in technical literature and models the capacitor or damper constitutive equations. These components are joined to each other to model the system much the same way a signal diagram is drawn. However, all connections carry power information rather than signal information. For a mechanical system power is a product of force and velocity, whereas power in an electrical system is voltage times current. The connections essentially are connecting components that exchange two types of physical quantities that make up power and these components vary from one domain to another. This is quite different from a signal flow diagram like in a traditional SIMULINK model. Thus, a physical model looks quite similar to the schematic diagrams that are drawn in literature to represent such physical systems and are therefore more familiar to the user than a signal flow diagram. Finally, the model developer need not derive the governing differential equations before putting a successful model together. This is a huge advantage since the important step of mathematical abstraction can be bypassed and the model developer can be productive with confidence in a hurry.

Simscape is a Matlab tool available within the Simulink suite that uses a physical modeling approach for the development of System models. It employs the Physical Network approach, which differs from the standard Simulink modeling approach and is particularly suited to simulating systems that consist of real physical components. Figure 1.6 shows a Simscape model of the DC motor, where an inductor or a damper or a resistor looks like what they represent and the connections exchange physically meaningful quantities such as voltage, current, torque, etc.

1.4   SOFTWARE

In this text we have used Simscape for all the model development discussion and simulation results that are reported. As mentioned before, Simscape is a Matlab/SIMULINK toolbox and is available with a standard license of Matlab. The text is divided into logical sections to tackle Mechanical, Electrical, and Magnetic domains in three separate chapters before combining them in multi-domain example models in the last two chapters. Effort was made to ensure that a first-time user of the tool can start with this book and be productive fairly quickly. Matlab/Simulink is a very well-established tool and has extensive on-line documentation libraries. The user will be able to answer a lot of their own questions beyond what is discussed here by referring to the Matlab libraries.

CHAPTER 2

Introduction to Simscape

2.1 PHYSICAL NETWORK APPROACH TO MODELING USING SIMSCAPE

Simscape is a toolbox available within the Simulink environment. It consists of a set of block libraries and simulation features for modeling physical systems. Simscape is based on the Physical Network approach, which differs from the standard Simulink modeling approach and is particularly suited for simulating systems that consist of real physical components. We discussed the broad differences between the two approaches in Chapter 1.

In Simulink models mathematical relationships are modeled using blocks connected by signal carrying links. The blocks perform specific mathematical functions. Simscape uses blocks that represent physical objects and these blocks replicate the constitutive relations that govern the behavior of the physical object represented by the block. A Simscape model is a network representation of the system under design, based on the Physical Network approach. Most often the network resembles typical schematics of the systems that the user is familiar with. Thus, an electric or magnetic circuit looks like a circuit seen in texts. The same is true for other domains. The exchange currency of these networks is power (or energy flow over time), i.e., the elements transmit power amongst each other through their points of entry or exit which are also known as ports. These connection ports are nondirectional. They are similar to physical connections between elements. Connecting Simscape blocks together is analogous to connecting real components, such as mass, spring, etc. Simscape diagrams essentially mimic the physical system layout. Just like real systems, flow directions need not be specified when connecting Simscape blocks. In this physical network approach the two variables that make up power are known as the through variable (TV) and across variable (AV).

The number of connection ports for each element is determined by the number of energy flows it exchanges with other elements in the system or the number of points through which power may enter or leave the element. For example, a permanent magnet DC motor is a two-port device with electric power coming in from one port and mechanical or rotational power leaving from the other port.

Energy flow or power is characterized by two variables. In Simscape modeling world these two variables are known as Through and Across. Usually, these are the variables whose product is the energy flow in watts. They are the basic variables. For example, the basic variables for mechanical translational systems are force and velocity, for mechanical rotational systems—torque

Table 2.1: Through and across variables in different domains

Physical Domain	Across Variable	Through Variable
Electrical	Voltage	Current
Hydraulic	Pressure	Flow rate
Magnetic	Magnetomotive force (mmf)	Flux
Mechanical rotational	Angular velocity	Torque
Mechanical translational	Translational velocity	Force
Thermal	Temperature	Heat flow

and angular velocity, for hydraulic systems—flow rate and pressure, for electrical systems—current and voltage.

2.1.1   VARIABLE TYPES

Through—Variables that are measured with a gauge connected in series to an element.
Across—Variables that are measured with a gauge connected in parallel to an element.

The Through and Across variables for different domains are listed in the Table 2.1.

Generally, the product of each pair of Across and Through variables associated with a domain is power (energy flow in watts). The exceptions are magnetic domain (where the product of mmf and flux is not power, but energy).

2.1.2   DIRECTION OF VARIABLES

The variables used in calculation are characterized by their magnitude and sign. The sign is the result of measurement orientation. The same variable can be positive or negative, depending on the polarity of a measurement gage. Simscape library has sensor elements for each domain to measure variables.

Elements with only two ports are characterized with one pair of variables, a Through variable and an Across variable. Since these variables are closely related, their orientation is defined with one direction. For example, if an element is oriented from port A to port B (Figure 2.1), it implies that the Through variable (TV) is positive if it "flows" from A to B, and the Across variable is determined as $AV = AV_A - AV_B$, where AV_A and AV_B are the element node potentials or, in other words, the values of this Across variable at ports A and B, respectively.

This approach to the direction of variables has the following benefits.

- Provides a simple and consistent way to determine whether an element is active or passive. Energy is one of the most important characteristics to be determined during simulation. If the variables direction, or sign, is determined as described above, their product (that is,

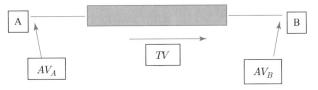

Figure 2.1: A generic element oriented from port A to port B.

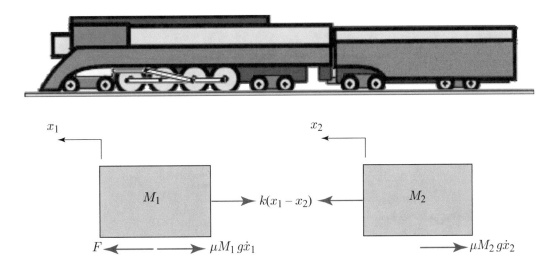

Figure 2.2: 2-mass train problem.

the energy flow) is positive if the element consumes energy, and is negative if it provides energy to a system. This rule is followed throughout the Simscape software.

- Simplifies the model description. Symbol $A \rightarrow B$ is enough to specify variable polarity for both the Across and the Through variables.

The following example illustrates a Physical Network representation of a simple 2-mass problem. In this example, we consider a toy train consisting of an engine and a car. Assuming that the train only travels in one dimension (along the track), the goal is to control the train so that it starts and stops smoothly, and so that it can track a constant speed command with minimal error in steady state (Figure 2.2).

The mass of the engine and the car are represented by M_1 and M_2, respectively. Furthermore, the engine and car are connected via a coupling with stiffness k. The coupling is modeled as a spring with a spring constant k. The force F represents the force generated between the wheels of the engine and the track, while μ represents the coefficient of rolling friction.

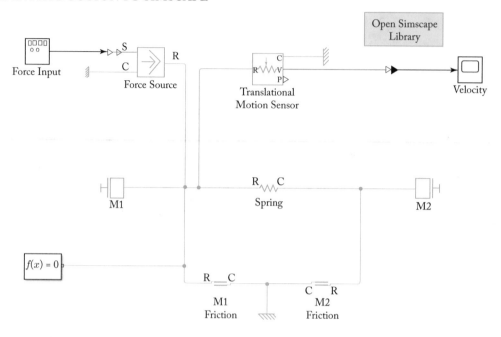

Figure 2.3: Simscape model of the 2-mass train problem.

Figure 2.2 shows the free body diagram of the system with the different forces acting on it. Using Newton's law and summation of forces for the two masses the governing equations may be written as:

$$M_1\ddot{x}_1 = F - k\,(x_1 - x_2) - \mu M_1 g \dot{x}_1$$
$$M_2\ddot{x}_2 = k\,(x_1 - x_2) - \mu M_2 g \dot{x}_2. \tag{2.1}$$

The corresponding Simscape model for the system is shown in Figure 2.3. This model is a fairly simplified version of the actual problem. However, the advantage of the networked system is that it is easily scalable; for example, a linear spring can be replaced by a nonlinear spring model or by a spring damper combination, if needed, without affecting any other parts of the model. The simple friction element can be swapped with a tire-road model if for example the two masses were representing a truck-trailer combination on a road rather than a train.

All the elements in a network are divided into active and passive elements, depending on whether they deliver energy to the system or dissipate/store it. Active elements (force and velocity sources, flow rate and pressure sources, etc.) must be oriented strictly in accordance with the line of action or function that they are expected to perform in the system, while passive elements (dampers, resistors, springs, pipelines, etc.) can be oriented either way.

In the 2-Mass train model the different elements, masses, the spring, and the friction elements are all shown to have two ports, R and C. All these are passive elements. The R and C

port connections of the passive elements could be reversed in the network without affecting any calculated results. For the Force source, an active element, the block positive direction is from port C to port R. In this block, port C is associated with the source reference point (ground), and port R is associated with the M_1. This means the force is positive if it acts in the direction from C to R, and causes bodies connected to port R to accelerate in the positive direction. The relative velocity is determined as $v = vC - vR$, where vR, vC are the absolute velocities at ports R and C, respectively, and it is negative if velocity at port R is greater than that at port C. The power generated by the source is computed as the product of force and velocity, and is positive if the source provides energy to the system. All this means that if the connections are reversed in the Force source or any other similar active elements, an opposite effect will be observed.

For more information on this check the block source or the block reference page on Simscape help pages if in doubt about the block orientation and direction of variables.

2.1.3   ELEMENT TYPES

All systems are made of a few basic components or elements that can be categorized into a few broad categories based on how they behave. It is important to understand these basic behaviors of elements to realize that there are plenty of similarities between systems in different domains. Within Simscape the element types are broadly categorized as Passive Elements and Active Elements. In a network all elements are either active and passive elements, depending on whether they deliver energy to the system or dissipate (or store) it.

2.1.4   STANDARD BEHAVIOR PASSIVE ELEMENTS

Among the many different passive elements that are available, element behavior can determine how they can be categorized. Here we have discussed the different categories.

Resistive Element: Energy Dissipating Device
The resistive element is a power dissipating device, i.e., the power that this element receives is lost from the system in the form of heat (most commonly) or in some other form of energy. These elements do not store energy. Figure 2.4 shows some examples of these types of devices in different domains. Some common resistance elements are electrical resistor, mechanical damper, wall roughness, or friction in a tube leading to pressure loss in a hydraulic circuit.

The constitutive relationships for these elements all look similar:

- for the electrical resistance this relationship is: *Voltage* $= R * (current)$; where R is the electrical resistance;

- for linear mechanical damping this relationship is: *Force* $= B * (velocity)$; where B is the damping coefficient;

- for rotational mechanical damping the relationship is: *Torque* $= B * \omega$; where B is the rotational damping coefficient; and

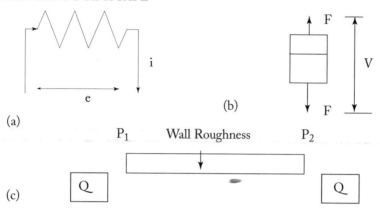

Figure 2.4: Examples of resistive elements: (a) electrical resistance, (b) mechanical damper, and (c) wall roughness.

Table 2.2: Resistance elements in different domains

	Domain Specific Relationship	Units for the Resistance Parameter	Power Dissipated
Mechanical Translation	$F = B\,v;\ v = F/B$	$B = N\text{-}s/m$	$Fv = F^2/B = v^2B$
Mechanical Rotation	$T = B\,\omega;\ \omega = T/B$	$B = N\text{-}m\text{-}s$	$T\omega = B\,\omega^2$
Electrical	$V = RI;\ I = V/R$	$R = V/A = \text{Ohms}$	$VI = V^2/R = I^2R$
Hydraulic	$P = RQ;\ Q = P/R$	$R = N\text{-}s/m^5$	$PQ = P^2/R = Q^2R$

- for the hydraulic resistance the equation is: $P = R * (volume flow rate)$; where R is hydraulic resistance.

For the resistive elements the rate of energy dissipated is the product of the across variable and the through variable. Thus, the power dissipated may be written as:

$$Power(t) = Through(t) * Across(t). \tag{2.2}$$

The specific relationships for power for all domains are listed in Table 2.2. Although the examples shown here are all linear relationships the Resistance relationship does not necessarily have to be linear. There are many practical situations where the resistance elements could have nonlinear relationships. Table 2.2 summarizes pertinent information about Resistance elements in different domains.

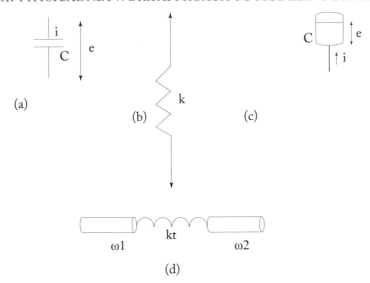

Figure 2.5: Capacitive elements: (a) linear electric capacitor, (b) mechanical spring, (c) fluid stored at some height or accumulator, and (d) torsional spring.

Capacitive Element: Energy Storage Device
The capacitive element is a storage device that stores and releases energy but does not dissipate it. The constitutive relationship for these elements are:

- for the electrical domain it is $V = \frac{Q}{C}$; i.e.,
  voltage = charge/capacitance;

- for the translational mechanical domain the relationship is: $F = k * x$;
  *force = spring constant * (displacement);*

- for the rotational mechanical domain the relationship is: $T = k_t * \theta$;
  *Torque = rotational spring constant * (angular displacement);*

- for the hydraulic domain the relationship is: $P = V/C$;
  pressure = volume/capacitance.

Figure 2.5 shows some examples of capacitive elements in different domains. The electrical capacitor and the mechanical spring are both capacitive elements. In the hydraulic domain the potential energy stored in a fluid that is stored in a tank at a height serves as a capacitive element. So does an accumulator in a hydraulic circuit.

Energy stored in these types of devices may be written as an integral of instantaneous power. And instantaneous power is equal to the product of instantaneous Across and Through

Table 2.3: Capacitive elements in different domains

	Domain Specific Relationship	Units for the Capacitance Parameter	Energy Stored
Mechanical Translation	$F = kx;\ x = F/k$	$k = N/m$	$E = \dfrac{1}{2}kx^2 = \dfrac{1}{2}\dfrac{F^2}{k}$
Mechanical Rotation	$T = k_t\theta;\ \theta = T/k_t$	$kt = N\text{–}m/rad$	$E = \dfrac{1}{2}k_t\theta^2 = \dfrac{1}{2}\dfrac{T^2}{k_t}$
Electrical	$V = q/C;\ q = VC$	$C = A\text{–}s/V = Farad(F)$	$E = \dfrac{1}{2}\dfrac{q^2}{C} = \dfrac{1}{2}CV^2$
Hydraulic	$P = V/C;\ V = PC$	$C = m^5/N$	$E = \dfrac{1}{2}\dfrac{V^2}{C} = \dfrac{1}{2}CP^2$

variables. The derivation below is written for the electrical specific quantities but is applicable for all domains:

$$
\begin{aligned}
Energy(t) &= \int Power(t)dt = \int e(t)f(t)dt = \int V(t)I(t)dt \\
&= \int V(t)\frac{dQ(t)}{dt}dt = \int V(t)C\frac{dV(t)}{dt}dt \\
&= \frac{1}{2}CV^2 = \frac{1}{2C}q^2.
\end{aligned}
\tag{2.3}
$$

Table 2.3 summarizes pertinent information about capacitive elements in different domains.

Inductor/Inertia: Energy Storage Device

An Inductor or Inertia element is a second type of energy storage device. Although both the capacitive and inductive elements are energy storage devices we have to consider them separately since the constitutive relationship (i.e., the nature of their behavior) are different as well as the type of energy they store. For example, a capacitor stores electrostatic energy and the inductor stores electromagnetic energy. Similarly, a spring stores elastic/potential energy whereas a mass or inertia stores kinetic energy.

For the inductor elements the constitutive relationship can be written as the following relationships for different domains:

- for the electrical domain: $\lambda = Li$;
  Flux linkage = (Inductance)(current);

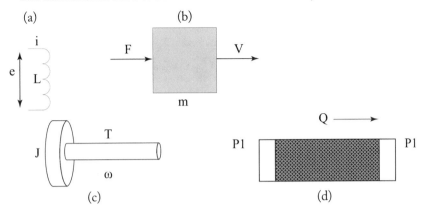

Figure 2.6: 1-Port inductive elements: (a) electrical inductor, (b) mechanical mass, (c) rotational inertia, and (d) fluid inertia.

- for the translational mechanical domain: $p = mv$;
  momentum = (*mass*)(*velocity*);

- for the rotational mechanical domain: $p = J\omega$;
  angular momentum = (*Polar moment of inertia*)(*angular velocity*); and

- for the hydraulic domain: $P_p = IQ$;
  pressure momentum = (*Hydraulic inertia*)(*volume flow rate*).

Some real components that behave as these types of elements are mass or inertia in the mechanical domain, electrical inductance of a coiled wire that carries a current, and the inertia of a mass of fluid that flows through conduit. Figure 2.6 shows schematics that represent these types of elements in the different domains.

The energy stored in these types of devices may be written as an integral of instantaneous power. And instantaneous power is equal to the product of instantaneous flow and effort variables. The derivation below is written for the electrical specific quantities but is applicable for all domains:

$$
\begin{aligned}
Energy(t) &= \int Power(t)dt = \int e(t)f(t)dt = \int V(t)I(t)dt \\
&= \int \frac{d\lambda(t)}{dt}I(t)dt = \int \frac{1}{L}\lambda(t)\frac{d\lambda(t)}{dt}dt \\
&= \frac{1}{2L}\lambda^2 = \frac{1}{2}LI^2.
\end{aligned}
\tag{2.4}
$$

Table 2.4 summarizes the inductor variables in different domains as well as the constitutive relationships.

Table 2.4: Inductor variables in different domains

	Domain Specific Relationship	Units for the Inductance Parameter	Energy
Mechanical Translation	$p = mv; \; v = p/m$	$m = kg = N\text{-}s^2/m$	$E = \dfrac{1}{2}mv^2 = \dfrac{1}{2}\dfrac{p^2}{m}$
Mechanical Rotation	$p_t = J\omega \; ; \; \omega = p_t/J$	$J = N\text{-}m\text{-}s^2$	$E = \dfrac{1}{2}J\omega^2 = \dfrac{1}{2}\dfrac{p_t^2}{J}$
Electrical	$\lambda = Li; \; i = \lambda/L$	$L = V\text{-}s/A = Henrys\;(H)$	$E = \dfrac{1}{2}Li^2 = \dfrac{1}{2}\dfrac{\lambda^2}{L}$
Hydraulic	$p_p = IQ; \; Q = p_p/I$	$I = N\text{-}s^2/m^5$	$E = \dfrac{1}{2}IQ^2 = \dfrac{1}{2}\dfrac{p_p^2}{I}$

Transformer

There are certain passive elements in a system which neither dissipate nor store energy in a system, but just transmit them from one part of the system to another. A transformer element is one type of such element for which two like-variables on the input and the output side are related to each other by a transformer factor in a way such that the input power and the output power are the same. If we call the two sides of a system that a transformer links as 1 and 2, respectively, the general relationship in a transformer element may be written as:

$$Through2\,(t) = N * Through1\,(t)$$
$$Across1\,(t) = N * Across2\,(t),$$

(2.5)

where N is the transformer factor. Simple manipulation of these two expressions will confirm that the input power and output power are equal.

Some examples of the transformer elements are: electrical transformer, levers, gear trains, hydraulic plunger-cylinder, rack and pinion, etc. Figure 2.7 shows schematics of some of the transformer elements. In each case the transformer factor may be obtained from a basic understanding of the system behavior. For example, in the electrical transformer the factor would be the ratio of the number of coils in both windings. The transformer factor for a lever would be the ratio of the lengths of the lever arms. For a gear train it could be the ratio of the gear teeth (or gear diameters), for a hydraulic ram it would be the area of the plunger (since the pressure multiplied by area is equal to the force).

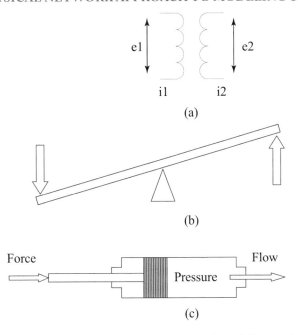

Figure 2.7: Schematics showing transformer elements within different domains: (a) electrical transformer, (b) mechanical transformer (lever), and (c) hydraulic transformer (plunger).

Gyrator

A gyrator element is similar to a transformer that it transmits power and does not store or dissipate it. The main difference is how the two key variables on the input side is related to the two key variables on the output side. For gyrators, it is such that the input through variable and the output across variable are related to each other by a factor and the input across variable and the output through variable are also related to each other by the same factor. These relationships also satisfy the requirement that the total input power is equal to the total output power.

The generic constitutive relationship for a gyrator may be expressed as:

$$Through2\,(t) = M * Across1\,(t)$$
$$Through1\,(t) = M * Across2\,(t). \tag{2.6}$$

One of the most common type of gyrator is the DC motor. The output torque of a DC motor is a linear function of the current in the motor armature and the back emf induced in the motor circuit is a function of the rotation in the shaft. Thus,

$$T\,(t) = (Torque\ constant) * i\,(t)$$
$$E\,(t) = (Torque\ constant) * \omega\,(t). \tag{2.7}$$

Another example of a gyrator is an electromagnet. For an electromagnet the relationships are:

$$Magnetomotive\ force\ (t) = (Number\ of\ coils) * Current\ (t)$$
$$Flux\ Linkage\ (t) = (Number\ of\ coils) * MagneticFlux\ (t). \qquad (2.8)$$

In each case the gyrator factor may be obtained from one's basic understanding of the system behavior. For example, in the DC motor the factor would be torque constant for the motor that is dependent on several factors including the magnetic flux density and motor geometry. For an electromagnet it is the number of coils in the electric circuit that helps to create the magnetic fields.

2.1.5   ACTIVE ELEMENTS

There are in general two types of sources that are used in these types of system models. Both these sources supply energy (or power) but in one case the power is supplied with a known through variable and in the other case the power is supplied with a known across variable. For example, a voltage source in an electrical system supplies power to an electrical circuit using a defined or known voltage profile. The current drawn is determined by the load of the system that is receiving the power. Similarly, a velocity source in a mechanical system supplies power with a known velocity profile and the force part of the power is determined by the system that is receiving power. All these are active elements and the direction of the voltage or velocity or force or any other quantity will be reversed with the reversing of polarity of these elements.

2.1.6   CONNECTOR PORTS AND CONNECTION LINES

Simscape blocks may have two types of ports: physical conserving ports and physical signal ports. They function differently and are connected in different ways. These ports and connections between them are described in detail below.

2.1.7   PHYSICAL CONSERVING PORTS

Simscape blocks have special conserving ports. Conserving ports are connected with physical connection lines, distinct from normal Simulink lines. Physical connection lines have no directionality and represent the exchange of energy flows/power, according to the Physical Network approach. Conserving ports can only be connected to other conserving ports of the same type (i.e., mechanical elements are connected by mechanical line types and electrical elements are connected by electrical line type, etc.). The physical connection lines that connect conserving ports together are nondirectional lines that carry physical variables (Across and Through variables) rather than signals. In passive and active element blocks in Simscape they are shown at C and R. Physical connection lines cannot be connected to Simulink ports or physical signal ports. In the example exercise shown in Figure 2.3, the connection lines connecting the masses with the spring and the friction elements are all Physical Conserving ports.

Two directly connected conserving ports must have the same values for all their Across variables (such as pressure or angular velocity). Branches can be added to physical connection lines. When branching happens, components directly connected with one another continue to share the same Across variables. Any Through variable (such as flow rate or torque) transferred along the physical connection line is divided among the components connected by the branches. This division is determined automatically by the dynamics of the system in consideration. For each Through variable, the sum of all its values flowing into a branch point equals the sum of all its values flowing out. This is similar to the Kirchoff's current law. For improved readability of block diagrams, each Simscape domain uses a distinct default color and line style for the connection lines. For more information, see Domain-Specific Line Styles in Simscape Help.

2.1.8   PHYSICAL SIGNAL PORTS

Physical signal ports carry signals between Simscape blocks. They are connected with regular connection lines, similar to Simulink signal connections. Physical signal ports are used in Simscape block diagrams instead of Simulink input and output ports to increase computation speed and avoid issues with algebraic loops. Physical signals can have units associated with them. The units along with the parameter values can be specified in the block dialogs, and Simscape software performs the necessary unit conversion operations when solving a physical network. Simscape Foundation library contains, among other sublibraries, a Physical Signals block library. These blocks perform math operations and other functions on physical signals, and allow you to graphically implement equations inside the Physical Network. Physical signal lines also have a distinct style and color in block diagrams, similar to physical connection lines in Simulink. In the model shown in Figure 2.3 the connection between the Force Input block to the S port of the force actuator is a physical signal connector. Similarly, the output from the velocity sensor that is connected to the V terminal (which is a physical signal port) is a physical signal connection. In both cases the two double arrows are used to transform a Simulink signal to a physical signal. These double-arrow symbols represent the **S_PS** (Signal to Physical Signal) or **PS_S** (Physical signal to Signal) transformation.

2.2   GETTING STARTED WITH SIMSCAPE

In the next few chapters many Simscape models and model development will be discussed with appropriate background information. Here we will just touch on how to get the model building process started. To open a new Simscape model type "**ssc_new**" in the MATLAB command window. That opens up a new model file and it looks like Figure 2.8.

The file opens with a few basic blocks already included. These are most common blocks that are used in any model. The **Scope** block would be fairly well known to Simulink users. This is used to graphically display any function (e.g., inputs or outputs from a simulation). The two sets of double arrows are the **PS_Simulink** and **Simulink_PS** blocks. These are the links between the traditional Simulink world where we deal with signals (with no physical meaning)

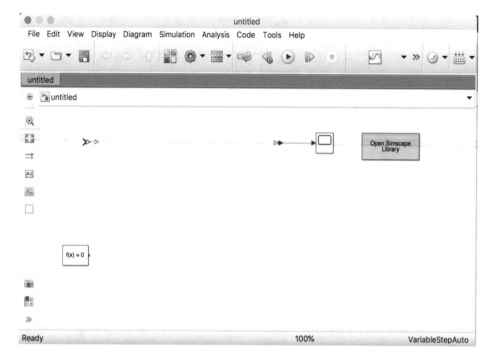

Figure 2.8: A new model file.

and the physical modeling world of Simscape where elements exchange physical variables. The Simulink_PS block changes a Simulink signal to a physical signal whose type and unit can be set by double clicking on the block. And the PS_Simulink block does the exact opposite, i.e., changes a physical signal such as velocity, voltage, displacement, etc., into a signal that can then be displayed by the scope. The block which shows **f(x) = 0** is known as the solver configuration block. This can be used to setup a local solver or a global solver. For every model, this block has to be attached some place in the model in order to carry out the simulation.

A Note About Solvers: The default solver is **ode 45**. Although this will work fine for most of the examples discussed here it is strongly recommended that you change the solver to a stiff solver (ode15s, ode23t, or ode14x). Do this by selecting "**Configuration Parameters**" from the Simulation menu, selecting the solver pane from the list on the left, and changing the "Solver" parameter to ode15s. Then click OK (Figure 2.9).

The use and need for all these blocks will be better clarified in the next few chapters in the contest of the examples that are discussed. The **Open Simscape Library** block can be used to get the top level of the Simscape Library opened. Clicking on this button opens the window in Figure 2.10.

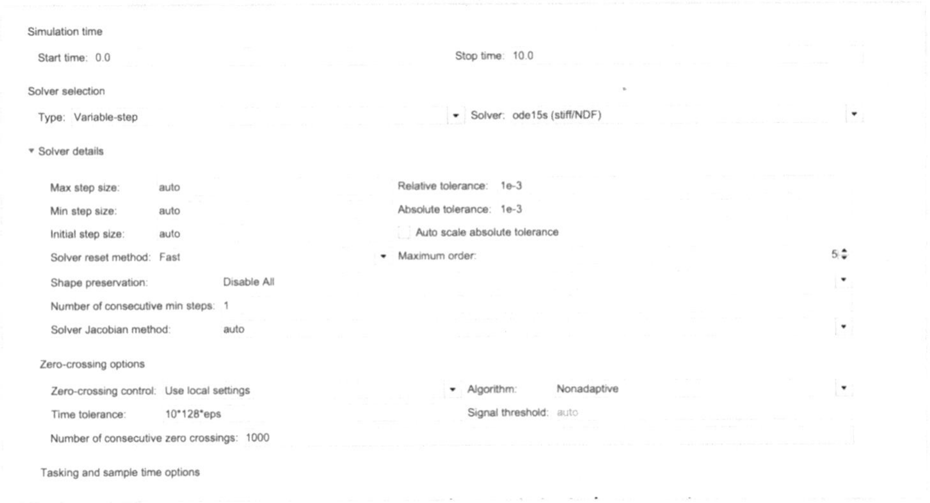

Simulation time
Start time: 0.0
Stop time: 10.0
Solver selection
Type: Variable-step
Solver: ode15s (stiff/NDF)
Solver details
Max step size: auto
Min step size: auto
Initial step size: auto
Solver reset method: Fast
Shape preservation: Disable All
Number of consecutive min steps: 1
Solver Jacobian method: auto
Relative tolerance: 1e-3
Absolute tolerance: 1e-3
Auto scale absolute tolerance
Maximum order: 5
Zero-crossing options
Zero-crossing control: Use local settings
Time tolerance: 10*128*eps
Number of consecutive zero crossings: 1000
Algorithm: Nonadaptive
Signal threshold: auto
Tasking and sample time options

Figure 2.9: Solver setting.

Library: simscape
File Edit View Display Diagram Analysis Help
simscape
simscape
Foundation Library
Utilities
Driveline
Electronics
Multibody
Power Systems
Simscape 4.2
Copyright 2006-2017 The MathWorks, Inc.
Ready
100%

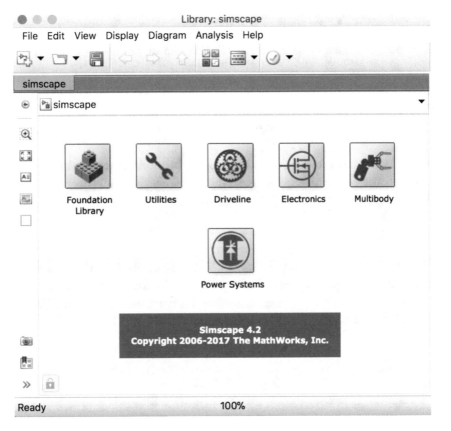

Figure 2.10: Simscape library.

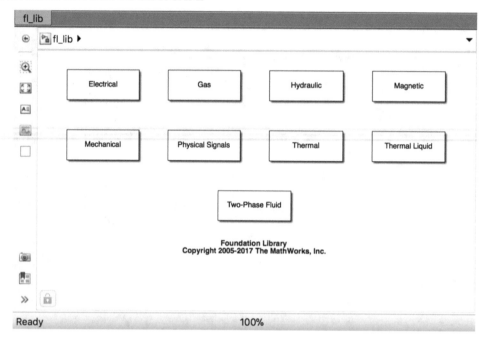

Figure 2.11: Available libraries within Simscape's Foundation library.

Along with the foundation library Simscape has several specialized libraries which are significantly well developed and advanced. In this book we will mostly focus on the Foundation library. Clicking on the Foundation library opens access to some of the basic libraries that are available.

Figure 2.11 shows all the basic domains that can be modeled using Simscape libraries. In this introductory text we have mostly discussed the use of Mechanical, Electrical, and Magnetic libraries. Those discussions start in with the next chapter.

2.3   SUMMARY

In this chapter we introduced the capabilities of Simscape toolbox of Matlab along with some basic concepts of Mechatronics. We also discussed the similarities among elements that are used in different domains to illustrate that there is a strong unifying concept among different domains even though we learn about them in very different contexts.

CHAPTER 3

Modeling Mechanical Translation and Rotation

3.1 INTRODUCTION

In this chapter we will introduce the readers to some of the basic elements in the Mechancial Systems library of Simscape and describe the development of some mechanical system models and their simulation. For the earlier examples in the text we will describe in detail every step of the model development process; but as the user gets familiar with Simscape model building we will rely on the knowledge gained by the user and focus mostly on the critical parts in our description.

3.1.1 SOME BASIC MECHANICAL SYSTEMS

Within the Foundation library of Simscape if we choose the Mechanical library we will find that there are several folders and each folder contain a number of building blocks for mechanical systems. In the next few paragraphs we have introduced those building blocks.

Mechanical Sensor: Four Sensor Blocks (Figure 3.1) for Linear and Rotational Motion
- Ideal Force Sensor—Force sensor in mechanical translational systems

- Ideal Rotational Motion Sensor—Motion sensor in mechanical rotational systems

- Ideal Torque Sensor—Torque sensor in mechanical rotational systems

- Ideal Translational Motion Sensor—Motion sensor in mechanical translational systems

Mechanical Sources: In the Source Folder there are Four Sources (Figure 3.2)
- Ideal Angular Velocity Source—Ideal angular velocity source in mechanical rotational systems

- Ideal Force Source—Ideal source of mechanical energy that generates force proportional to the input signal

- Ideal Torque Source—Ideal source of mechanical energy that generates torque proportional to the input signal

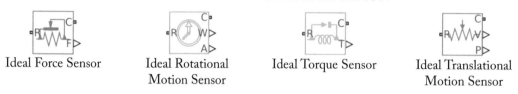

| Ideal Force Sensor | Ideal Rotational Motion Sensor | Ideal Torque Sensor | Ideal Translational Motion Sensor |

Figure 3.1: Sensors used in mechanical systems.

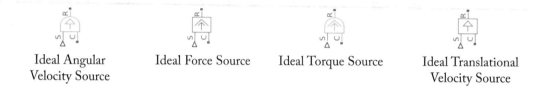

| Ideal Angular Velocity Source | Ideal Force Source | Ideal Torque Source | Ideal Translational Velocity Source |

Figure 3.2: Sources for mechanical systems.

- Ideal Translational Velocity Source—Ideal velocity source in mechanical translational systems

Linear Mechanical Motion: In the Linear Motion Folder there are Eight Elements (Figure 3.3)
- Mass—Ideal mechanical translational mass

- Mechanical Translational Reference—Reference connection for mechanical translational ports

- Translational Damper—Viscous damper in mechanical translational systems

- Translational Free End—Translational port terminator with zero force

- Translational Friction—Friction in contact between moving bodies

- Translational Hard Stop—Double-sided translational hard stop

- Translational Inerter—Two-port inertia in mechanical translational systems

- Translational Spring—Ideal spring in mechanical translational systems

Angular Mechanical Motion: In the Rotational Motion Folder there are Eight Elements (Figure 3.4)
- Inertia—Ideal mechanical rotational inertia

- Mechanical Rotational Reference—Reference connection for mechanical rotational ports

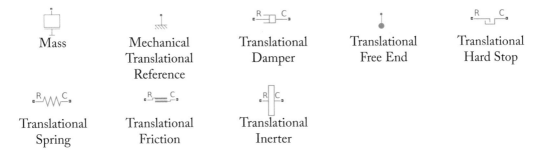

Figure 3.3: Mechanical elements for linear motions.

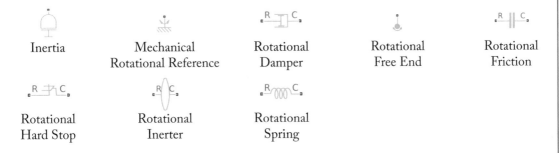

Figure 3.4: Mechanical elements for rotational motion.

- Rotational Damper—Viscous damper in mechanical rotational systems
- Rotational Free End—Rotational port terminator with zero torque
- Rotational Friction—Friction in contact between rotating bodies
- Rotational Hard Stop—Double-sided rotational hard stop
- Rotational Inerter—Two-port inertia in mechanical rotational systems
- Rotational Spring—Ideal spring in mechanical rotational systems

Mechanisms: In the Mechanism Folder there are Three Elements (Figure 3.5)
- Gear Box—Gear box in mechanical systems, transforms rotation to rotation
- Lever—Generic mechanical lever; transforms translation to translation
- Wheel and Axle—Wheel and axle mechanism in mechanical systems, transforms rotation to translation, and vice versa

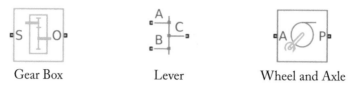

| Gear Box | Lever | Wheel and Axle |

Figure 3.5: Mechanism elements.

3.2   EXAMPLES

In the next few sections we have discussed development of mechanical models using Simscape. The steps followed in building the models are described along with the system response. Also, in the first example the implementation of a feedback control loop to obtain a desired response from the model is also discussed.

Example 3.1   Automotive Suspension

Some of the most common elements in any mechanical translational system are the spring, mass, and dampers. And the most common practical example is a system that includes these elements is an automotive suspension. In this example we have considered suspension for a vehicle. When the suspension system is designed, a 1/4 model (one of the four wheels) is used to simplify the problem to a 1-D multiple spring-damper system. Figure 3.6 shows a schematic for such a suspension for a bus. The two masses represent the vehicle mass and the wheel mass/suspension mass, respectively. A good automotive suspension system should have satisfactory road holding ability, while still providing comfort when riding over bumps and holes in the road. When the vehicle is experiencing any road disturbance (i.e., pot holes, cracks, and uneven pavement), the vehicle body should not have large oscillations, and the oscillations should dissipate quickly. Since the distance $X1 - W$ is very difficult to measure, and the deformation of the tire $(X2 - W)$ is negligible, we will use the distance $X1 - X2$ instead of $X1 - W$ as the output in our problem. Keep in mind that this is an estimation.

The road disturbance (W) in this problem will be simulated by a step input. This step could represent the vehicle coming out of a pothole. We will also discuss the design of a feedback controller so that the output $(X1 - X2)$ has an overshoot less than 5% and a settling time shorter than 5 s. For example, when the vehicle runs onto a 10-cm high step, the vehicle body will oscillate within a range of ± 5 mm and return to a smooth ride within 5 s.

The system parameters used are as follows.

1. (m1) body mass 2500 kg

2. (m2) suspension mass 320 kg

3. (k1) spring constant of suspension system 80,000 N/m

4. (k2) spring constant of wheel and tire 500,000 N/m

Model of Bus Suspension System (¼ Bus)

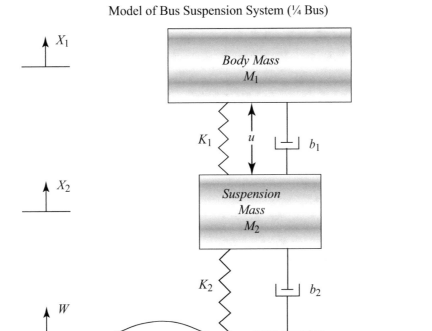

Figure 3.6: Schematic for a suspension system.

5. (b1) damping constant of suspension system 350 Ns/m

6. (b2) damping constant of wheel and tire 15,020 Ns/m

7. (u) control force = force from the controller we are going to design

Open a new Simscape model by typing `ssc_new` in the MATLAB command window. A new model opens up with a few commonly used blocks already in the model (Figure 3.7).

Assemble System to Represent Tire and Road Inputs
Below are the blocks to add. Here are some tips for adding blocks. These will be handy for all the models used in this text.

Tips for adding model elements

1. Use Quick Insert to add the blocks. Click in the diagram and type the name of the block. A list of blocks will appear and you can select the block you want from the list. Alternatively, the Open Simscape Library block can be used to look though the library of all blocks and pick the appropriate one.

f(x) = 0

Figure 3.7: A new model window.

2. After the block is entered, a prompt will appear for you to enter the parameter. Enter the variable names as shown in examples.

3. To rotate a block or flip blocks, right-click on the block and select Flip Block or Rotate block from the Rotate and Flip menu.

4. To show the parameter below the block name, see Set Block Annotation Properties in the Simulink documentation.

* Mechanical Translational Reference

* Ideal Translational Velocity Source

* Translational Spring   (Spring rate = k2)

* Translational Damper   (Damping coefficient = b2)

* Mass   (Mass = m2)

* Pulse Generator   (Amplitude = 1)

For the Pulse Generator, double-click on the block and set Period to 100, Pulse Width to 0.1.

Connect them as shown in Figure 3.8. Connect the output of the Pulse Generator to the Simulink-PS Converter block that is already in the diagram in the upper left corner (two triangles with a Simulink input), and connect the output of that block to the Ideal Translational Velocity Source.

You can give the blocks meaningful names by clicking on the block, then clicking on the name and editing it.

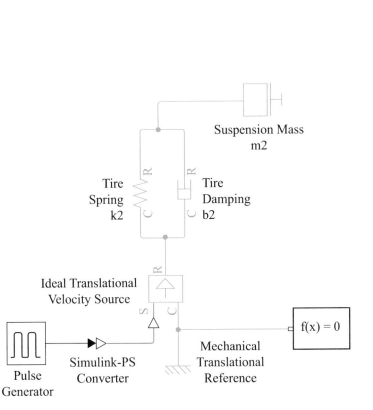

Figure 3.8: Building the first set of spring, mass, and damper system.

Sensing Key Values and Simulate

The following steps will display simulation results on the Scope.

1. Connect the R port of the Ideal Translational Motion Sensor as shown in Figure 3.9.

2. Make a copy of the Mechanical Translational Reference block to connect to the C port of the sensor.

3. Connect the "v" port of the sensor to the PS-Simulink converter already in the diagram (white triangle and black triangle which is connected to the Scope).

4. Make a copy of the PS-Simulink converter, and connect the "p" port to the Scope via the converter block as shown in Figure 3.9.

Before running the simulation, set the maximum step size to 1e-2.

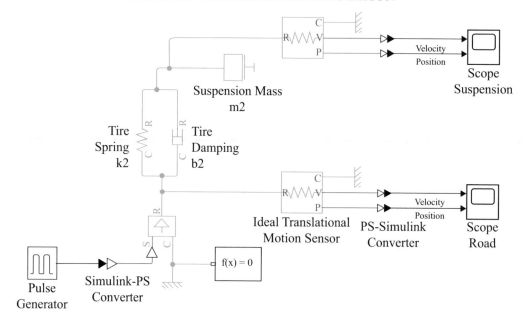

Figure 3.9: After adding the sensors and scopes.

1. Type CTRL-E to open the Configuration Parameters dialog (or choose from the Simulation menu in the menu bar).

2. Click on the Solver Pane.

3. Expand the Solver details section.

4. Set "Max step size" to 1e-2.

5. In Matlab workspace assign numerical values for all the variables, k2 = 500000, b2 = 15020, m2 = 320.

Run a simulation (type CTRL-T or press the green arrow run button). Figure 3.10 shows the road input, a 0.1 m bump and Figure 3.11 shows the velocity and position of the suspension mass.

Define Vehicle System

You can copy and paste the spring, damper, mass, and sensor blocks to model the vehicle as a second mass-spring-damper. Click and drag to select the blocks, then right-click on one of the selected blocks and drag to copy them (Figure 3.12).

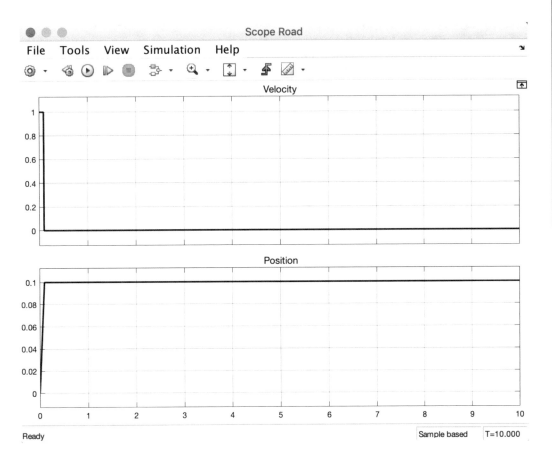

Figure 3.10: Road input profile velocity and position.

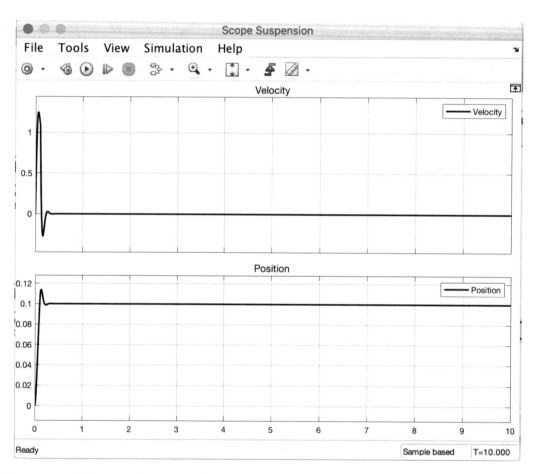

Figure 3.11: Velocity and position of suspension mass.

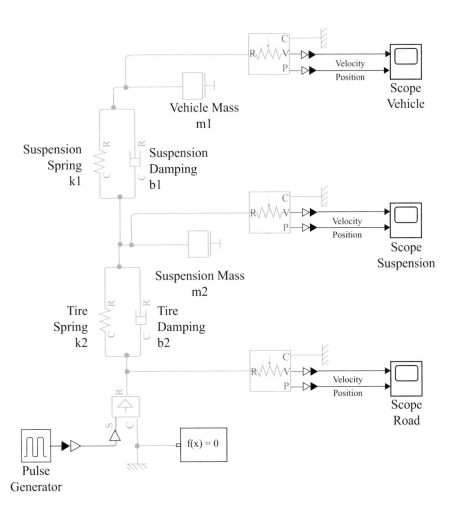

Figure 3.12: Suspension system model.

You will need to set the parameters to m1, b1, and k1, and we recommend giving the blocks meaningful names. Enter the values of m1, b1 and k1 by typing them in the Matlab workspace.

Run a simulation (type CTRL-T or press the green arrow). You will see the vehicle body oscillates for a long time. Set the stop time to 50 s to see how long it oscillates (Figure 3.13).

Define Control Input and Measure Additional States

At this stage we will implement a full state-feedback controller. The controller will use five state outputs, namely:

$$\left[X1 \; \frac{dX1}{dt} \; Y1 = (X1 - X2) \; \frac{dY1}{dt} \; \int Y1 \, dt \right]^{T}.$$

The controller uses the following feedback gain matrix:

$$K = [0 \; 2.3E6 \; 5E8 \; 0 \; 8E6].$$

Add an Ideal Force Source as shown in Figure 3.14 so that our control system can apply a force to quickly damp out the oscillations from the bump.

To measure the suspension deflection, add two Subtract blocks to calculate the difference between the relative positions and velocities of the vehicle and the suspension. Add an integrator to integrate the difference in relative motion of the vehicle and suspension. Connect them as shown in Figure 3.14.

Create Closed-Loop Control

To place the suspension model in a subsystem.

1. Replace the scopes with Out1 blocks and the two inputs with In1 blocks.

2. Click-and-drag to select all blocks except the Step, Pulse Generator, Scope Vehicle, and Scope Suspension Deflection.

3. Type CTRL-G (or right click and choose) to create the subsystem.

4. Resize the subsystem so that all ports can be seen.

The outputs of this system are the states that will be used by the controller. We wish to label them and order them the same as used for the control design steps. Rename and re-order the outputs as shown in the screenshot shown in Figure 3.15.

Back at the top level, delete the output connections and connect the subsystem outputs to a Mux block as shown in Figure 3.16. When you add the Mux block, you can set the parameter Number of inputs to 5. The output of the Mux block is the state vector that will be used by the controller.

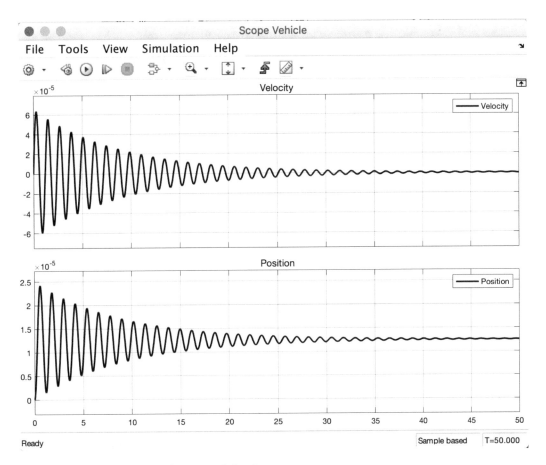

Figure 3.13: Vehicle body velocity and displacement.

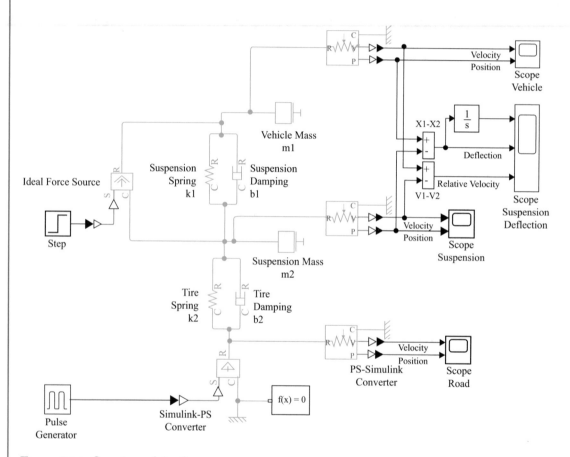

Figure 3.14: Creating of the five state outputs.

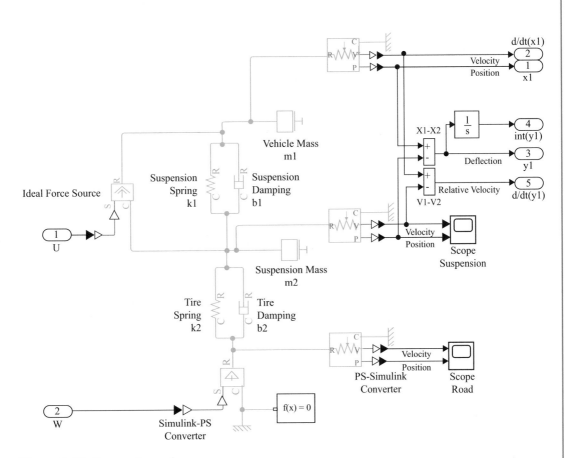

Figure 3.15: Suspension sub-system.

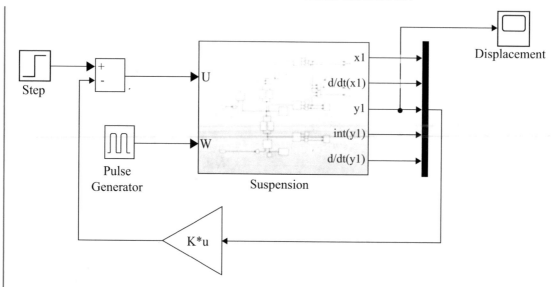

Figure 3.16: Closed-loop suspension model.

To use the control gains, add a Gain block and set parameter Gain to K. Double-click on the block and set the Multiplication parameter to Matrix (K*u). Connect the output of the Mux block to the input of the Gain block. The output of this block will be subtracted from the command signal to generate the actuator command.

Add a Subtract block, and connect the Gain and Step blocks as shown in Figure 3.16. Connect the output of the subtract block to the input that connects to the Ideal Force Source in the suspension subsystem.

Before running the simulation, set the inputs to model the case of the bus going over a bump.

1. In the Step block, set the Final value parameter to 0.

2. In the Pulse Generator block, set the Amplitude to 1.

3. In the workspace type the K value to be the five element gain matrix mentioned earlier.

Run the simulation (type CTRL-T or press the green arrow run button). The Scope shows that the displacement of the vehicle is much smaller than before due to the control system. You can set the stop time to 2 s to see the oscillations in more detail. The output (X1-X2) has a settling time less than 5 s and an overshoot less than 5%. For example, when the bus runs onto a 10-cm high step, the bus body will oscillate within a range of ±5 mm and will stop oscillating within 5 s. So the desired controlled behavior is achieved (Figure 3.17).

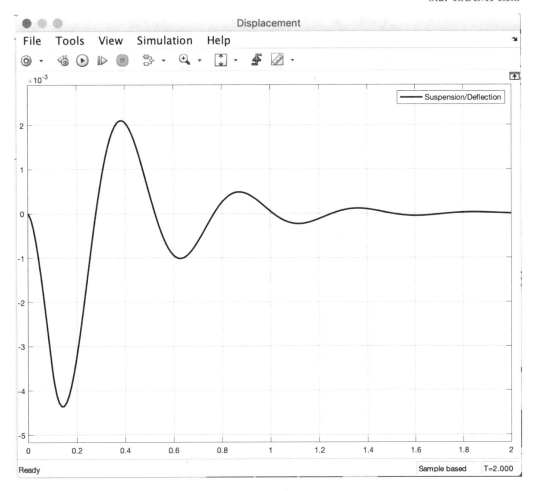

Figure 3.17: Controlled response of suspension.

Example 3.2   Half Car model

The study of dynamics of vehicles is a study of the dynamics of a 3-D rigid body.

In a 3-D model, there are six degrees of freedom—translations in three mutually perpendicular directions x, y, z and Φ, θ, ψ, rotations about the three axes. These three rotations are also called pitch, yaw, and roll, respectively (Figure 3.18). A complete vehicle dynamics model needs to consider all these motions since they are not all independent of each other. However, simplified 1-D and 2-D models are used extensively for many useful insights into system behavior. The previous example discussed is a 1-D model of a vehicle suspension. 1-D models are also called quarter car models since they represent one fourth of a car. Results obtained from the 1-D model only relate to one dimension, vertical displacement or heave motion. 2-D models

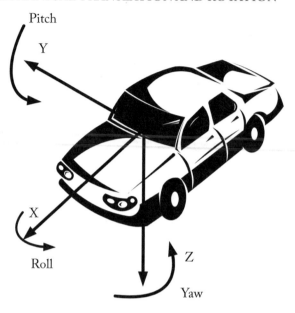

Figure 3.18: Pitch, yaw, and roll motion directions of a vehicle.

are a little bit more complex and are called half-car models. These models could account for the difference between the front of the car and the rear, location of the vehicle centroid, multiple inputs from the two wheels, etc. Thus, these models can provide more information on vehicle suspension behavior as well as the interaction of the front and rear suspensions. The output from the half-car model are two degrees of freedom, the vertical displacement or the heave, and rotation about y-axis or the pitch.

Figures 3.19 and 3.20 show the vehicle schematics including the center of mass and the corresponding motion directions for heave and pitch motions. The level of complexity to be included in the model can be decided by the user. The simpler approach is to include the vehicle mass along with the suspension at the front and rear represented through a pair of spring and damper elements. If more complexity is desired the wheel/suspension masses and the corresponding tire stiffness could be included as well (just like in the previous example). In this Half-car model the wheel masses and tire elastic properties are not included, although they can be easily added in the current modeling framework. There are 2 degrees of freedom in the 2-D model: (a) Heave and (b) Pitch.

Linked to these two motions there are two inertia (kinetic energy storage) elements, the vehicle mass and rotational inertia about the y-axis.

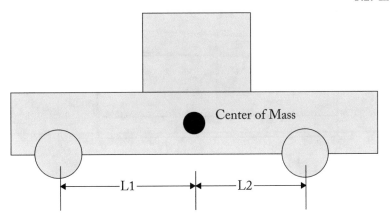

Figure 3.19: Schematic of half-car model.

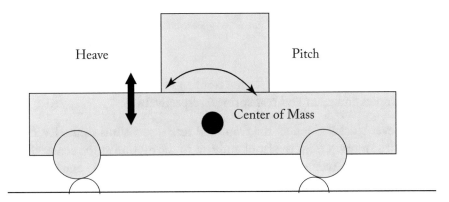

Figure 3.20: Half-car model showing heave and pitch direction motion.

Tips for adding model elements

1. Use Quick Insert to add the blocks. Click in the diagram and type the name of the block. A list of blocks will appear and you can select the block you want from the list. Alternatively, the Open Simscape Library block can be used to look though the library of all blocks and pick the appropriate one.

2. After the block is entered, a prompt will appear for you to enter the parameter. Enter the variable names as shown below.

3. To rotate a block or flip blocks, right-click on the block and select Flip Block or Rotate block from the Rotate and Flip menu.

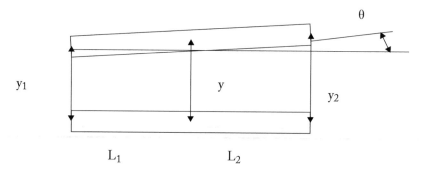

Figure 3.21: Transformer parameter calculations.

4. To show the parameter below the block name, see Set Block Annotation Properties in the Simulink documentation.

Model Development Procedure

1. Add a pair of Spring and Dampers for the front and the rear suspension. Name them front and rear dampers and front and rear springs, respectively.

2. Add two Lever blocks. Connect the front and rear suspensions to the lever block that is used to link the motion from the front and rear to the motion at the center of mass. There are two motions at the center of mass, the pitch and the heave. Use one lever for each type of motion. For both levers connect the suspensions to nodes A and B of the lever.

3. For the first lever, used in heave motion tracking, connect a Mass block to node C.

4. For the second lever, used in pitch motion tracking, connect a Wheel and axel block that will transfer the linear motion to a rotational motion and then connect an Inertia block to node C.

5. Attach two Velocity input blocks to the bottom ends of the front and rear suspensions.

6. Using the S-PS blocks add two Pulse generators from the Simulink library. We will use these to provide the different road inputs to the front and back. Refer to Figure 3.22.

Transformer Parameter Lever Relation Calculation

To determine the transformer parameters the vehicle is approximated as a rectangular piece and two of its positions are shown in Figure 3.21 to show the movement of three different locations of the vehicle, namely the point of attachment of the front and the rear suspensions and the center of mass.

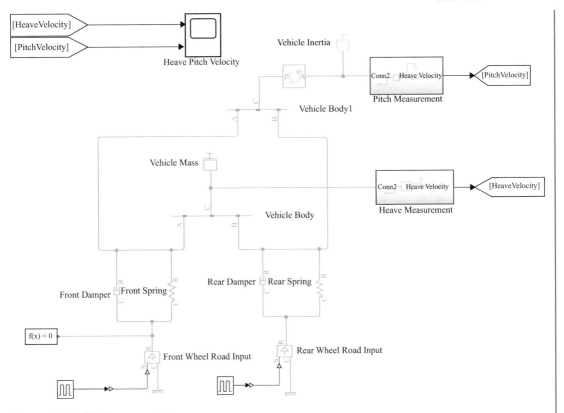

Figure 3.22: Half-car model.

$L1 + L2$ is the total distance between the front and back wheel.
θ is the rotational angle (pitch).
y is the vertical displacement of the center of the mass.

$$\theta = \frac{y2 - y1}{L1 + L2}$$
$$\dot{\theta} = \frac{\dot{y}2}{L1 + L2} - \frac{\dot{y}1}{L1 + L2} \qquad (3.1)$$
$$\dot{\theta} = \frac{\dot{y}2 - \dot{y}1}{L1 + L2}.$$

Therefore, $1/(L1 + L2)$ is the factor that transfers the linear motion at the two wheels into the rotational motion of the body. So the wheel and axel used will use $L1 + L2$ as the parameter for the wheel radius that will transfer the linear motion to angular motion. This relationship relates linear velocities into angular velocities.

The vertical displacement of the center of mass can be written as:

$$\dot{y} = \dot{y}1 + \frac{\dot{y}2 - \dot{y}1}{L1 + L2} \cdot L1$$

$$\dot{y} = \frac{\dot{y}1(L1 + L2) + \dot{y}2L1 - \dot{y}1L1}{L1 + L2}$$

$$\dot{y} = \frac{\dot{y}1L2 + \dot{y}2L1}{L1 + L2}$$

$$\dot{y} = \dot{y}1\left(\frac{L2}{L1 + L2}\right) + \dot{y}2\left(\frac{L1}{L1 + L2}\right).$$

(3.2)

This relationship of motion from the front and rear transmission to the motion of the vehicle centroid is achieved by providing $L1$ and $L2$ and dimensions of AC and BC of the lever that is connected to the mass.

In the model presented here the following parameters for a vehicle are used:

- Vehicle mass (sprung mass) = 1,500 kg

- Moment of Inertia about the CG for pitch motion = 2,160 kgm^2

- K1 (suspension, front) = 35,000 N/m

- K2 (suspension, rear) = 38,000 N/m

- B1 (damping coefficient, Front) = 1,000 Ns/m

- B2 (damping coefficient, Rear) = 1,100 Ns/m

- Distance between front axel and center of gravity (AC), $L1 = 1.4$ m

- Distance between rear axel and center of gravity (BC), $L2 = 1.7$ m

- Wheel radius used in the wheel and axel block ($AB + BC$) = $L1 + L2 = 3.1$ m

- The Simscape model is shown in Figure 3.22.

To avoid clutter in the model Goto and From blocks are used to store and use data for plotting. In this model they store the pitch and heave velocities. Two velocity sensors are used to measure the heave velocity and the pitch velocity. To reduce clutter in the model the sensor system is put into subsystem blocks. The heave measuring subsystem is shown in Figure 3.23.

At both the wheels a pulse input is used to simulate the behavior of the system running over a bump. The pulse generator on the front wheel has a magnitude of 1 unit at the front wheel and the pulse generator has a magnitude of 0.0 unit at the rear wheel. The front wheel pulse input is shown in Figure 3.24. The 1 unit road disturbance signifies a velocity input of 1 m/s. The pulse applies after a 1 s phase delay. The simulation is run for 10 s. This means that

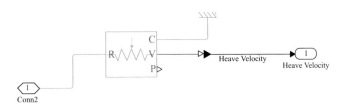

Figure 3.23: Heave measurement subsystem.

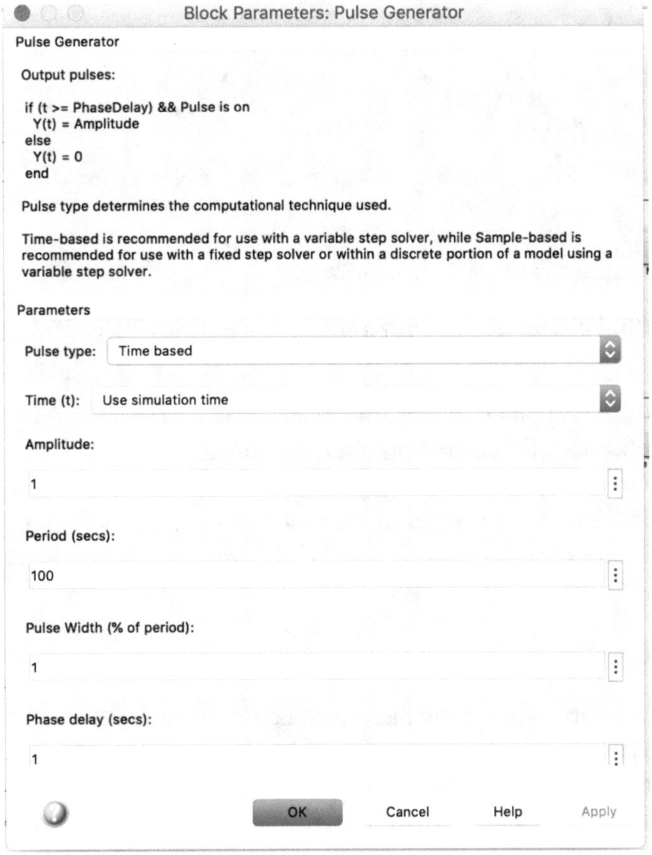

Figure 3.24: Input parameters for the first road disturbance.

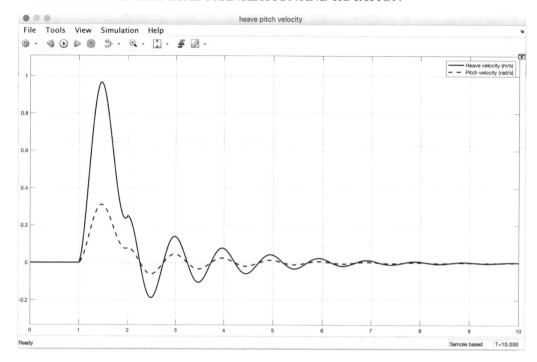

Figure 3.25: Heave and pitch velocity with a single road-disturbance.

the pulse of disturbance is applied on the front wheel after 1 s from the start. Figure 3.25 shows the velocity profiles for both the heave and pitch motions.

The model is now slightly altered by turning on a second road input on the rear wheel. The disturbance parameters are identical to the first input except the input comes with a phase delay of 4 s (the first input was after a delay of 1 s). Figure 3.26 shows the system response.

3.3 SUMMARY

In this chapter we have introduced the basic mechanical elements available in the foundation library and described two examples where some of the basic elements are used to model common mechanical systems. Since this is the first chapter in the book where model development is being introduced, we have followed a step-by-step approach. This will help the user get familiar with the software as she/he is also putting together a working model. In later chapters we hope to build on this basic understanding and describe model development with fewer instructions.

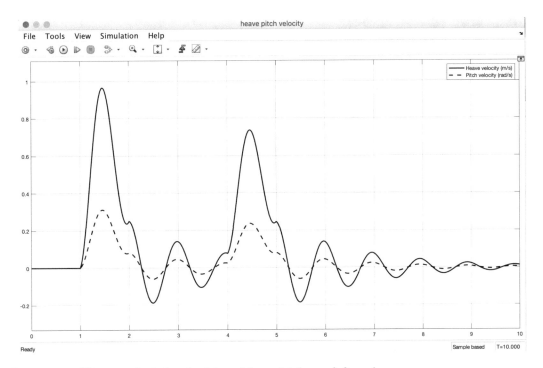

Figure 3.26: Heave and pitch velocities with multiple road disturbances.

CHAPTER 4

Modeling Electrical Systems

4.1   INTRODUCTION

In this chapter we will introduce the readers to some of the basic elements in the Electrical Systems library of Simscape and discuss the development of some electrical system models and their simulation.

4.1.1   SOME BASIC ELECTRICAL ELEMENTS

Within the Foundation library of Simscape choose the Electrical library to find several folders and each folder contain a number of building blocks for electrical systems. All the blocks are introduced here.

Electrical Sensors: Two Electrical Sensors (Figure 4.1)
- Current Sensor—Current sensor in electrical systems

- Voltage Sensor—Voltage sensor in electrical systems

Electrical Sources: Ten Electrical Sources (Figure 4.2)
- AC Current Source—Ideal sinusoidal current source

- AC Voltage Source—Ideal constant voltage source

- Controlled Current Source—Ideal current source driven by input signal

- Controlled Voltage Source—Ideal voltage source driven by input signal

- Current-Controlled Current Source—Linear current-controlled current source

- Current-Controlled Voltage Source—Linear current-controlled voltage source

- DC Current Source—Ideal constant current source

- DC Voltage Source—Ideal constant voltage source

- Voltage-Controlled Current Source—Linear voltage-controlled current source

- Voltage-Controlled Voltage Source—Linear voltage-controlled voltage source

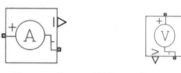

Current Sensor Voltage Sensor

Figure 4.1: Electrical sensors.

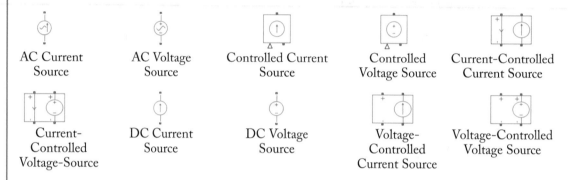

| AC Current Source | AC Voltage Source | Controlled Current Source | Controlled Voltage Source | Current-Controlled Current Source |

| Current-Controlled Voltage-Source | DC Current Source | DC Voltage Source | Voltage-Controlled Current Source | Voltage-Controlled Voltage Source |

Figure 4.2: Electrical sources.

Electrical Elements: Seventeen Electrical Elements (Figure 4.3)

- Capacitor—Linear capacitor in electrical systems

- Diode—Piecewise linear diode in electrical systems

- Electrical Reference—Connection to electrical ground

- Gyrator—Ideal gyrator in electrical systems

- Ideal Transformer—Ideal transformer in electrical systems

- Inductor—Linear inductor in electrical systems

- Infinite Resistance—Electrical element for setting initial voltage difference between two nodes

- Memristor—Ideal memristor with nonlinear dopant drift approach

- Mutual Inductor—Mutual inductor in electrical systems

- Op-Amp—Ideal operational amplifier

- Open Circuit—Electrical port terminator that draws no current

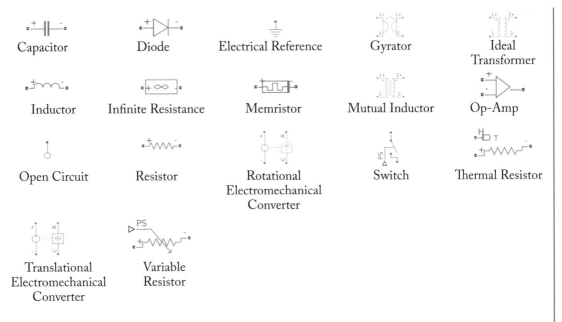

Figure 4.3: Electrical elements.

- Resistor—Linear resistor in electrical systems

- Rotational Electromechanical Converter—Interface between electrical and mechanical rotational domains

- Switch— Switch controlled by external physical signal

- Thermal Resistor—Resistor with thermal port

- Translational Electromechanical Converter—Interface between electrical and mechanical translational domains

- Variable Resistor—Linear variable resistor in electrical systems

4.2   EXAMPLES

We have discussed several examples from the electrical domain. Here are a set of instructions to help with adding new elements in a system model.

Tips for adding model elements

1. Use Quick Insert to add the blocks. Click in the diagram and type the name of the block. A list of blocks will appear and you can select the block you want from the list. Alternatively, the Open Simscape Library block can be used to look though the library of all blocks and pick the appropriate one.

2. After the block is entered, a prompt will appear for you to enter the parameter. Enter the variable names as shown in the examples.

3. To rotate a block or flip blocks, right-click on the block and select Flip Block or Rotate block from the Rotate and Flip menu.

4. To show the parameter below the block name, see Set Block Annotation Properties in the Simulink documentation.

Example 4.1   RL Circuit

An RL circuit is a first-order system; it has only one storage device, the inductor. Applying Kirchhoff's voltage law an electrical circuit with a voltage source, a resistance and an inductance can be easily found to be governed by first order differential equation:

$$L\frac{di}{dt} + Ri = v(t). \tag{4.1}$$

The solution of first order differential equations are an exponential function of the form $e^{-t/\tau}$, where τ is called the time constant and its specific value is determined by system parameters. The solution for this particular equation is:

$$i = \frac{v}{R}\left(1 - e^{-t/(\frac{L}{R})}\right). \tag{4.2}$$

So the time constant is L/R. For a current that is growing from an initial value of 0 it will take one time constant to rise by 63.2% of the final value and by 4 times the time constant the current reaches over 99% of the final value. The final value of the current in this circuit is v/R, i.e., in the long term (steady state) the inductor behaves like a conductor.

The model is developed by adding appropriate blocks as follows.

1. Type ssc_new in the Matlab command window to open a new model file.

2. Select and place the Resistor, Inductor, and Electrical Reference elements in the workspace.

3. Select and place the DC Voltage Source element in the workspace.

4. Select and place the Current Sensor element in the workspace.

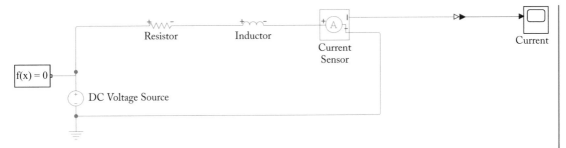

Figure 4.4: RL circuit model.

5. Select and connect PS-Simulink Converter to the Scope block. This block converts the physical signal (PS) to a unit-less Simulink output signal. Connect its input to the upper output port of the current sensor. This is the I port, where I stands for "current." The I port outputs the current as a physical signal which has units. The other ports ("+" and "−") are physical connections to the rest of the circuit.

6. The Solver Configuration block defines the solver settings for this Simscape physical network. The Simulink solver for the entire model must be set separately. For this example, do not change any of the parameters in this block (all three boxes should be unchecked).

7. Assign parameters for the components: $R = 5$ Ohms, $L = 0.04$ H, and $v = 10$ V.

8. Connect the elements as shown in Figure 4.4.

The time constant for this system is L/R. If v is constant, the steady state current will be v/R. For our values this gives $L/R = 0.008$ s and a steady state current of 2 Amps, which is reached to within 2% after $4(0.008) = 0.032$ s. Set the simulation time to 0.06 s and run the model. The results (Figure 4.5) agree with our analytical discussion, the current reaches near steady state of 2 A in a little over 0.03 s.

Example 4.2   RLC Circuit
The simplest form of an RLC circuit (a circuit with a Resistor, Inductor, and Capacitor) consists of the three elements connected with a voltage source. Using Kirchhoff's voltage law the differential equation for a series RLC circuit can be written as:

$$L\ddot{q} + R\dot{q} + \frac{q}{C} = V(t), \tag{4.3}$$

where L, R, and C are values of the Inductance, Resistance, and Capacitance, respectively; q is the charge in the capacitor, its first derivative is the current and the second derivative is the rate of change of current, and $V(t)$ is the applied voltage. This form of the governing equation is a

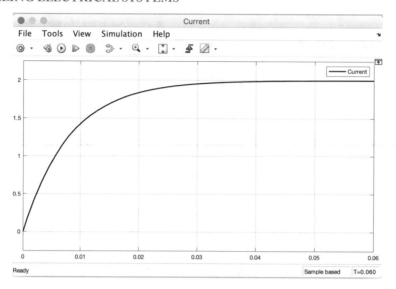

Figure 4.5: RL circuit current.

classic second order equation. It is also analogous to the second order equation in a mechanical system that consists of a mass, damper and spring:

$$m\ddot{x} + B\dot{x} + kx = F(t).\tag{4.4}$$

The solution for this non-homogenous equation has two parts, the Complementary Function (CF) which is the solution of the homogenous equation and the particular integral (PI) that is the solution for a specific non-homogenous term:

$$q(t) = q(t)_{CF} + q(t)_{PI}.\tag{4.5}$$

For obtaining the CF the standard practice is to assume the solution as a standard exponential form, Ae^{st}.

This results in the expression:

$$\left(Ls^2 + Rs + \frac{1}{C}\right)Ae^{st} = 0.\tag{4.6}$$

This leads to two roots of the characteristic equation (the quadratic equation in s). And the nature of those roots dictates if the system will be underdamped, overdamped, or critically damped. The roots can be derived as:

$$s_1 = -\zeta\omega_n + \omega_n\sqrt{\zeta^2 - 1}$$
$$s_2 = -\zeta\omega_n - \omega_n\sqrt{\zeta^2 - 1},\tag{4.7}$$

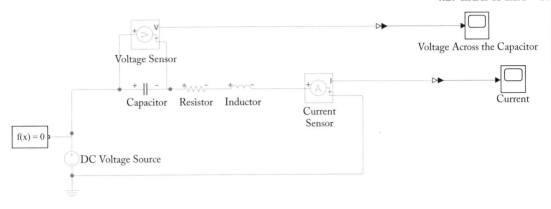

Figure 4.6: RLC model.

where

$$\omega_n = \frac{1}{\sqrt{LC}}, \qquad \zeta = \frac{R}{2}\sqrt{\frac{C}{L}}. \tag{4.8}$$

ω_n is the natural frequency of the system and ζ is the damping ratio. For a mechanical system, L is equivalent to mass, m, and C is equivalent to $1/k$, and R is equivalent to B, the damping coefficient. Swapping the electrical variables with the mechanical variables lead to corresponding expressions for natural frequency and damping ratio in a mechanical system:

$$\omega_n = \frac{\sqrt{k}}{\sqrt{m}}, \qquad \zeta = \frac{B}{2\sqrt{km}}. \tag{4.9}$$

A damping ratio < 1 leads to an underdamped or oscillatory system and a damping ratio equal to or greater than 1 leads to a non-oscillatory system. This means that the CF part of the total solution has two characteristic parameters, the natural frequency and the damping ratio. The form of the Particular Integral, the second part of the total solution, mirrors the nature of the non-homogenous part of the equation $V(t)$.

The model of the RLC circuit can be easily built by modifying the RL circuit model with the inclusion of a capacitor element and a voltage sensor is added to track the voltage across the capacitor. The parameters used in this model are: $C = 1$ F, $L = 0.01$ H, $R = 0.1$ Ohm, and $V = 10$ V. The model is shown in Figure 4.6. The model is simulated for 10 s.

For this set of parameters the natural frequency is 10 rad/s and the damping ratio is 0.5. And because $V(t)$ is a constant voltage (10 V) the solution shown in Figures 4.7 and 4.8 show some oscillation in the transient region and a steady value of voltage and current in the long-term.

The model is modified with new parameters: $C = 0.1$ F, $L = 0.01$ H, $R = 0.1$ Ohm, $V = 10$ V. This time the natural frequency is 31.67 rad/s and damping ratio is 0.158. The sys-

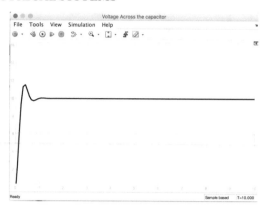

Figure 4.7: Voltage across the capacitor.

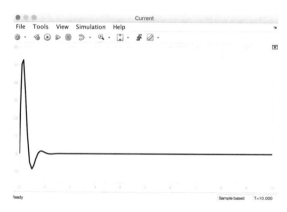

Figure 4.8: Current in the circuit.

tem response shown in Figures 4.9 and 4.10 show a higher frequency of oscillation and lower damping than the previous case. But in the steady state the system converges to a steady value.

The model is modified again with parameters C, R, and L unchanged but a waveform is used for the applied voltage instead of a constant value. Natural frequency is still 31.67 rad/s and damping ratio is 0.158. The model is shown in Figure 4.11. The pulse generator details is shown in Figure 4.12. The system response shown in Figures 4.13 and 4.14 clearly depicts that the result combines the oscillatory part, CF, superposed on a pulse response that is the PI.

Example 4.3   Basic Electronic Circuits: Integrator

Operational amplifiers are sometimes called the workhorse of electronic applications. Operational amplifiers (Op-Amps) form the core of many signal conditioning applications such as amplifiers, filters, integrators, differentiators, etc. Using the same basic Op-Amp and with

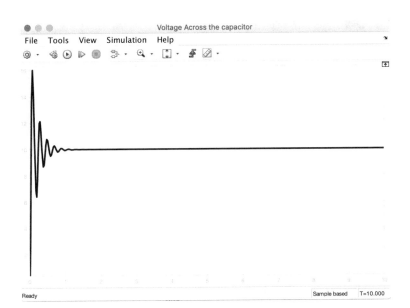

Figure 4.9: Voltage across the capacitor with increased natural frequency and lower damping.

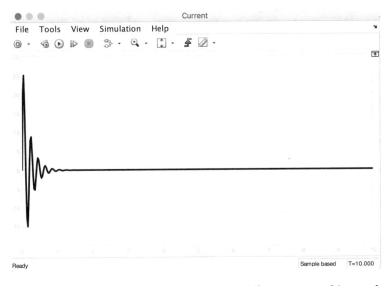

Figure 4.10: Current in the circuit with increased natural frequency and lower damping.

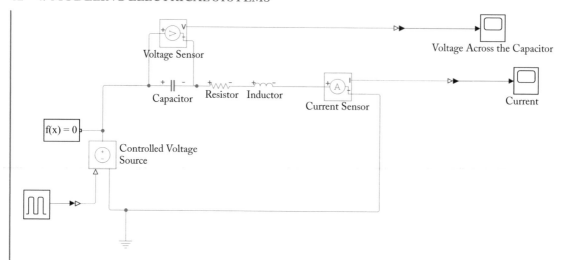

Figure 4.11: RLC model for variable voltage input.

proper combination of resistors, capacitors, and feedback loops a large number of applications of Op-Amps can be implemented. The Op-Amp is an integrated circuit that consists of many electronic components. Overall behavior of Op-Amps is characterized by a very high input resistance and a very low output resistance. In the open circuit mode (although Op-Amps are never used in this mode) they can provide voltage amplification of the order of millions. By using proper external circuit elements (resistors, capacitors, inductors, etc.) the user can control this amplification factor to almost any desired value as well as alter the behavior of the circuit to create many devices. Here we present one such device circuit, an integrator. The integrator uses a resistor in series with an input voltage at the positive terminal of the Op-Amp and a feedback capacitor. It has been shown that the output voltage is given by the relationship:

$$v_{out} = \frac{-1}{RC} \int v_{in} dt. \tag{4.10}$$

The output voltage is not only an integral of the input voltage but is multiplied by a factor of $1/RC$ and its sign is reversed.

The following steps may be followed to develop the model.

1. Type ssc_new in the Matlab command window to open a new model file.

2. Select and place the Resistor, Capacitor, Op-Amp, Pulse generator, and Electrical Reference elements in the workspace.

3. Select and place the Controlled Voltage Source element in the workspace.

4. Select and place the Voltage Sensor element in the workspace.

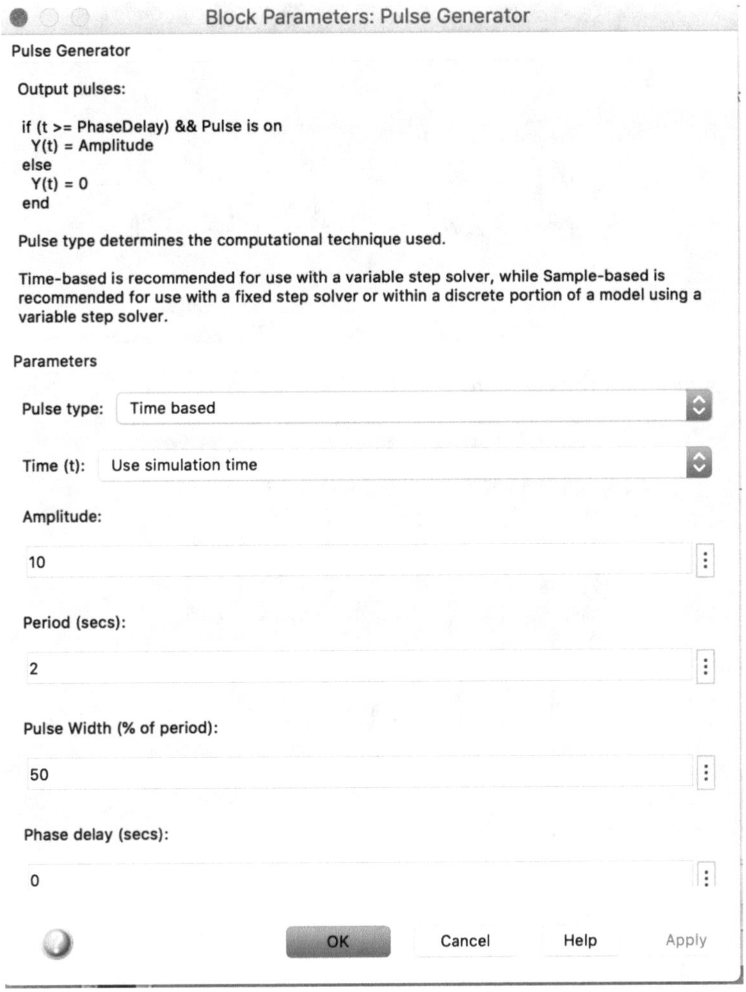

Figure 4.12: Pulse generator details.

5. Select and connect PS-Simulink Converter to the Scope block. This block converts the physical signal (PS) to a unit-less Simulink output signal. Connect its input to the upper output port of the voltage sensor. This is the V port, where V stands for "voltage." The V port outputs the voltage as a physical signal which has units. The other ports ("+" and "−") are physical connections to the rest of the circuit.

6. The Solver Configuration block defines the solver settings for this Simscape physical network. The global default solver settings are used for this model. So leave all boxes unchecked in the solver settings window.

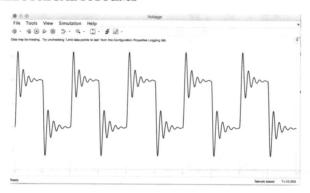

Figure 4.13: Voltage across the capacitor.

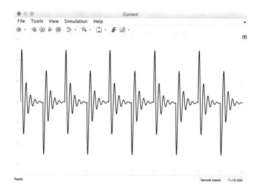

Figure 4.14: Current in the circuit.

7. Assign parameters for the components: $R = 100$ Ohms, $C = 0.01$ F.

8. Connect the elements as shown in Figure 4.15.

9. Set up the pulse generator input as shown in Figure 4.16 to model the pulse input voltage of a maximum value of 10 V.

Simulation of the model provides output that is shown in Figure 4.17. As expected, the integrated output voltage grows with time and is reversed in direction from the input voltage.

Example 4.4   Basic Electronic Circuits: Full Wave Bridge Rectifier
This example deals with another device circuit that is commonly used in many applications to convert AC input into DC outputs. This circuit is called a rectifier and uses an electronic component, the diode. The diode is essentially an electronic valve that allows flow of current in one direction but stops current from flowing in the opposite direction.

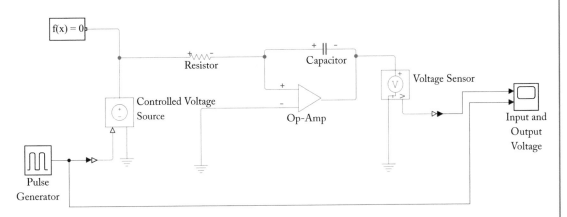

f(x) = 0

Resistor

Capacitor

Voltage Sensor

Controlled Voltage
Source

Op-Amp

Input and
Output
Voltage

Pulse
Generator

Figure 4.15: **Integrator model.**

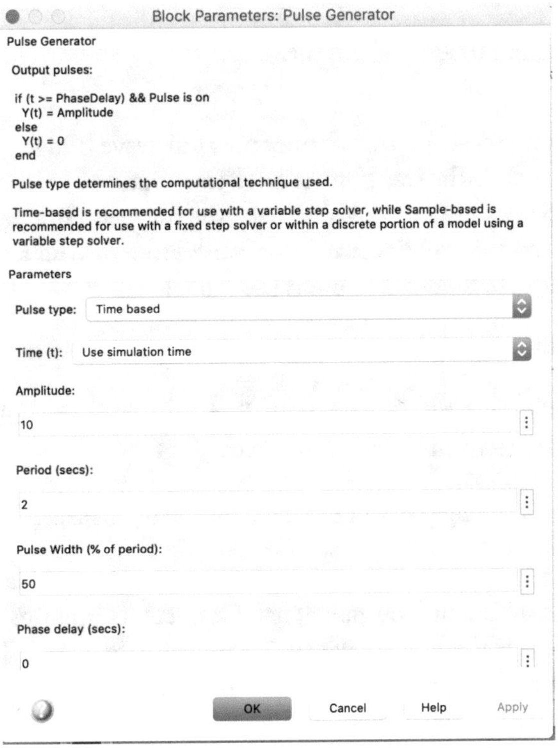

Figure 4.16: **Input voltage setting.**

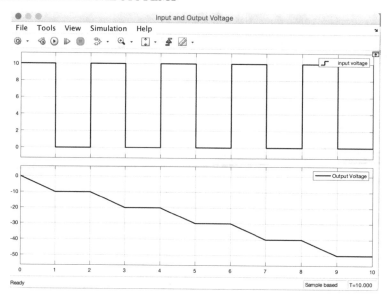

Figure 4.17: Input pulse and integrated output.

This example uses an ideal AC transformer plus full-wave bridge rectifier. It first converts 120 Volts AC to 12 Volts DC. The transformer has a turns ratio of 14, stepping the supply down to 8.6 Volts rms, i.e., $8.6 * sqrt(2) = 12$ Volts peak. The full-wave bridge rectifier plus capacitor combination then converts this to DC. A resistor is used to represent a typical load.

The following steps are followed to build the model.

1. Type ssc_new in the Matlab command window to open a new model file.

2. Select and place the Resistor, Capacitor, 4 diodes, Transformer, AC voltage source, and Electrical Reference elements in the workspace.

3. Select and place two Voltage Sensor elements in the workspace.

4. Select and connect two PS-Simulink Converters to the Scope blocks. This block converts the PS to a unit-less Simulink output signal. Connect its input to the outputs of the two voltage sensors. This is the V port, where V stands for "voltage." The V port outputs the voltage as a physical signal which has units. The other ports ("+" and "−") are physical connections to the rest of the circuit.

5. The Solver Configuration block defines the solver settings for this Simscape physical network. The global default solver settings are used for this model. So leave all boxes unchecked in the solver settings window.

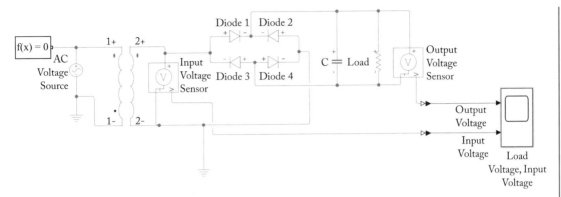

Figure 4.18: Full-wave rectifier.

6. Assign parameters for the components: $R = 100$ Ohms, $C = 470 \ \mu F$, transformer winding ratio $= 14$ and default settings for diode including forward voltage of 0.6 V. Set the input AC voltage frequency to 60 Hz and peak voltage to 120 sqrt(2).

7. Connect the elements as shown in Figure 4.18. Set simulation time to 0.05 s and run simulation.

Figure 4.19 shows both the input sinusoidal voltage as well as the output rectified voltage. The DC output voltage has some ripples but is near-steady at 10 V. This model can be used to size the capacitor required for a specified load. For a given size of capacitor, as the load resistance is increased, the ripple in the DC voltage increases. This model is re-run by reducing the capacitor size by an order of 10 to, $C = 47 \ \mu F$ and the simulation is re-run. The results are shown in Figure 4.20. It is quite clear that the ripples have increased significantly. The model is rerun one more time with $C = 4,700 \ \mu F$ a 10 times higher value than the original C. Simulation results are shown in Figure 4.21. Ripples have gone down even more than the original but because of the higher C the rise to the steady voltage is slow.

4.3   SUMMARY

In this chapter we discussed model development for electrical systems using some of the basic components in the foundation library that relates to electrical and electronic systems.

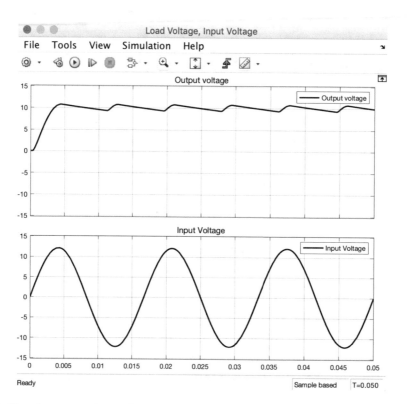

Figure 4.19: Simulation results with $C = 470 \ \mu F$.

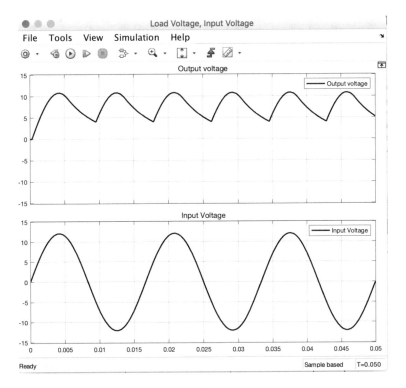

Figure 4.20: Simulation results with $C = 47\ \mu$F.

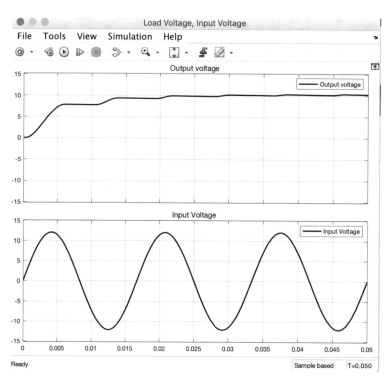

Figure 4.21: Simulation results with $C = 4,700\ \mu\text{F}$.

CHAPTER 5

Modeling Magnetic Systems

5.1 INTRODUCTION

We will begin this chapter with a quick discussion of some of some of the basic concepts and components associated with magnetic circuits.

5.1.1 MAGNETIC CIRCUITS: BACKGROUND

Magnetic circuits are somewhat similar to electric circuits. The basic physical quantities in a magnetic circuit are magnetic field intensity, magnetomotive force, magnetic flux density, etc. are critical quantities in a magnetic system. Let us quickly define some of these basic quantities for our understanding of any magnetic system.

Magnetic Field Intensity, H

If a current flows through a straight conductor, a magnetic field is induced around the conductor. This field is quantified by magnetic field intensity, H (A/m). It is defined as the current in the conductor divided by the length of the closed path around the conductor (Amperes Law):

$$\oint_{closed\ path} H d\ell = \sum i. \tag{5.1}$$

If the path is a circular path this equation can be algebraically (Figure 5.1) expressed as:

$$H = \frac{i}{2\pi r}, \tag{5.2}$$

where r is the radius of the closed path. This means that the further one moves away from the conductor H reduces. Also, H is not a function of the material/medium that surrounds the conductor, i.e., it is independent of material properties.

Magnetomotive Force, Ni

Extending the idea of H, if there are multiple conductors inside a closed path (Figure 5.1), the algebraic sum of all the currents flowing through all the conductors determines what the H will be. This is one of the most powerful concepts that has been used extensively in the design of electromagnetic devices:

$$H = \frac{\sum i}{l}. \tag{5.3}$$

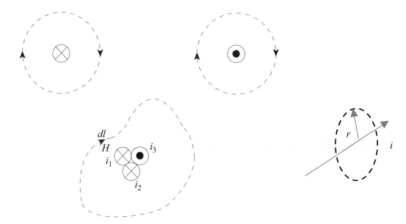

Figure 5.1: Magnetic field intensity for a closed path around current carrying conductors.

Figure 5.2: Closed path enclosing a number of coils in a conductor (only coil cross-section shown).

If we consider a coil of wire of N number of turns (Figure 5.2) with a current i passing through it, and a closed path that engulfs the coil, the H becomes:

$$H = \frac{Ni}{l}. \tag{5.4}$$

Electrical machines such as motors, transformers, solenoids, etc., will have a large number of coils of wire. This design takes advantage of the concept of generating a high H by passing a small current through a large number of coils. This quantity, Ni, in the numerator of the H equation is called the magnetomotive force (Ampere turns). This is analogous to the electromotive force in an electrical circuit, i.e., just like the voltage difference drives a current through an electrical circuit, the magnetomotive force drives flux through a magnetic circuit.

Magnetic Flux Density, B

The magnetic field intensity, H, causes magnetic lines of forces, flux, in the medium surrounding the source of H. The magnetic flux density, B, is related to H through the following relationship:

$$B = \mu_r \mu_0 H, \tag{5.5}$$

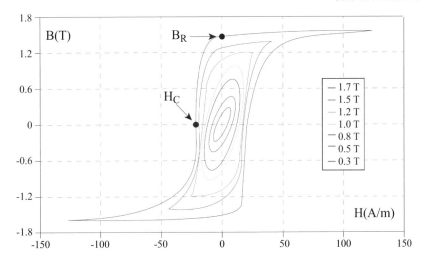

Figure 5.3: Sample hysteresis curves relating $B - H$ relationship for magnetic materials.

where μ_0 is the *permeability of free space*, $4\pi \times (10)^{-7}$ Tm/A, and μ_r is the *relative permeability of the material*; μ_r is 1 for air or vacuum, and of the order of 10s to 100s of thousand for magnetic steel and other good magnetic materials.

Even though this $B - H$ relationship as stated seems to be linear, in reality $B - H$ relationship is highly nonlinear. The true relationship between B and H is usually expressed through a set of hysteresis curves and essentially the magnetic material cycles through these curves as H increases and decreases (due to increasing and changing direction of the current). Figure 5.3 shows a typical hysteresis curve. Although the overall relationship of $B - H$ is nonlinear, for certain practical ranges of operation the $B - H$ relationship can be approximated linear, as stated in the above equation.

Magnetic Flux
Magnetic flux or lines of forces is defined as the total number of lines of force passing through a certain area and can be easily expressed as:

$$\phi = (A_c)B, \tag{5.6}$$

where A_c is the cross-section of the medium through which the lines of force are passing.

Figure 5.4 shows a typical magnetic circuit where the mmf, Ni, is generated by a current passing through an electrical coil. This mmf drives a flux through the magnetic circuit or closed path of magnetic material through which this flux travels. Thus, a magnetic circuit is analogous to an electric circuit. If we combine some of the basic equations discussed before in the context

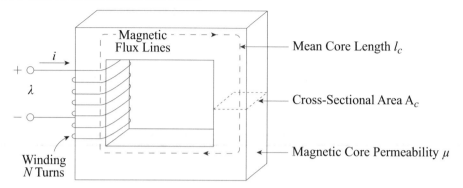

Figure 5.4: A typical magnetic circuit.

of the magnetic circuit we can write the following:

$$\phi = (A_c)\, B = (A_c)\, B = A_c \mu_0 \mu_r H = (A_c \mu_0 \mu_r) \frac{Ni}{l_c} \tag{5.7}$$

$$Ni = \left(\frac{l_c}{A_c \mu_0 \mu_r} \right) \phi \tag{5.8}$$

$$Ni = \left(\frac{l_c}{A_c \mu_0 \mu_r} \right) \phi = \Re\phi. \tag{5.9}$$

$\Re$ is analogous to resistance in an electrical circuit and is called Reluctance. Ni is magnetomotive force and is analogous to the voltage/source in an electrical circuit and the flux ϕ is analogous to the current in the electrical circuit. In Simscape, flux is the through variable and magnetomotive force is the across variable.

5.2   SOME BASIC MAGNETIC ELEMENTS

Magnetic Sources: There are Four Magnetic Forces (Figure 5.5)
- Controlled Flux Source—Ideal flux source driven by input signal
- Controlled MMF Source—Ideal magnetomotive force source driven by input signal
- Flux Source—Ideal flux source
- MMF Source—Ideal magnetomotive force source

Magnetic Sensors: There are Two Magnetic Sensors (Figure 5.6)
- Flux Sensor—Ideal flux sensor
- MMF Sensor—Ideal magnetomotive force sensor

Controlled Flux   Controlled MMF   Flux Source   MMF Source
Source   Source

Figure 5.5: Magnetic sources.

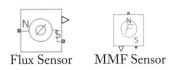

Flux Sensor   MMF Sensor

Figure 5.6: Magnetic sensors.

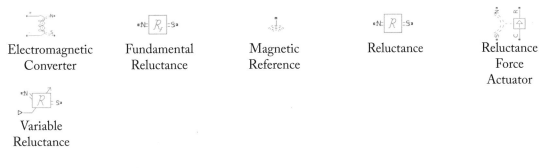

Electromagnetic   Fundamental   Magnetic   Reluctance   Reluctance
Converter   Reluctance   Reference   Force
                                                               Actuator

Variable
Reluctance

Figure 5.7: Magnetic elements.

Basic Magnetic Elements: There are Six Basic Magnetic Elements (Figure 5.7)

- Electromagnetic Converter—Lossless electromagnetic energy conversion device

- Fundamental Reluctance—Simplified implementation of magnetic reluctance

- Magnetic Reference—Reference connection for magnetic ports

- Reluctance—Magnetic reluctance

- Reluctance Force Actuator—Magnetomotive device based on reluctance force

- Variable Reluctance—Variable reluctance in electromagnetic systems

5.3   EXAMPLES

Here we consider several examples from the magnetic domain. In constructing the models we will be adding different elements in the model. Here are some standard tips that will be useful for developing all the models.

Tips for adding model elements

1. Use Quick Insert to add the blocks. Click in the diagram and type the name of the block. A list of blocks will appear and you can select the block you want from the list. Alternatively, the Open Simscape Library block can be used to look though the library of all blocks and pick the appropriate one.

2. After the block is entered, a prompt will appear for you to enter the parameter. Enter the variable names as shown in the examples.

3. To rotate a block or flip blocks, right-click on the block and select Flip Block or Rotate block from the Rotate and Flip menu.

4. To show the parameter below the block name, see Set Block Annotation Properties in the Simulink documentation.

Example 5.1 Simple Magnetic Circuit

This first example is a very simple magnetic circuit. It has one mmf source and a single reluctance. This circuit is very similar to the one showed in Figure 5.4.

1. Type ssc_new in the Matlab command window to open a new model file.

2. Select and place the Fundamental Reluctance, MMF Source, and Magnetic Reference elements from the Simscape Magnetics Libraries.

3. Select and place the Flux Sensor element.

4. Set the MMF value to 1,000 and the reluctance to 80.

5. Connect the elements as shown in Figure 5.8.

6. Leave the Solver setting at "auto" for this simple simulation.

7. Run the simulation.

   The simulation result shows a steady flux value in this magnetic circuit (Figure 5.9).

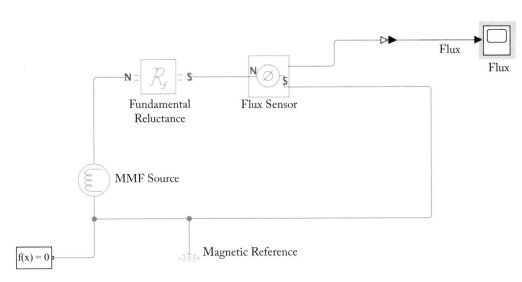

Figure 5.8: A simple magnetic circuit.

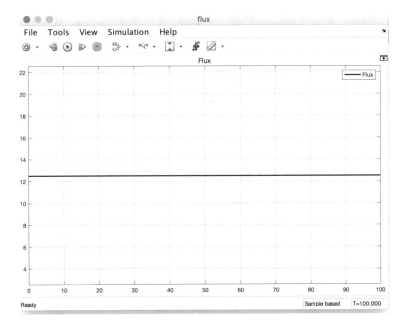

Figure 5.9: Magnetic flux in the circuit as measured by the flux sensor.

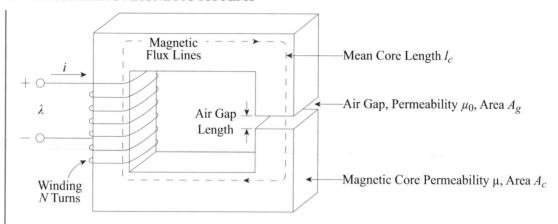

Figure 5.10: A schematic of a magnetic circuit with an airgap.

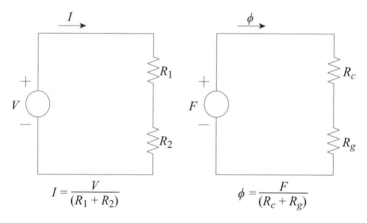

Figure 5.11: A magnetic circuit with two reluctances and its equivalency with an electric circuit.

Example 5.2   Magnetic Circuit with Airgap

Airgaps are a core feature in many magnetic circuits. Motors and generators have small airgaps between the stator and the rotor to allow physical motion of one part of the magnetic circuit while the other part remains stationary. Even for a very small airgap the reluctance of the airgap ends up being several orders of magnitude higher than the reluctance of the iron core. This model incorporates the reluctance due to a small airgap in the magnetic circuit (Figure 5.10). Using an approach similar to an electrical circuit, the Reluctance of the magnetic circuit in the presence of the airgap becomes a sum of two reluctances, one for the magnetic core (a relatively small value) and one for the airgap (a relatively larger value). The two reluctances are in series in the circuit. In this model the two reluctances are separately accounted for in the magnetic circuit.

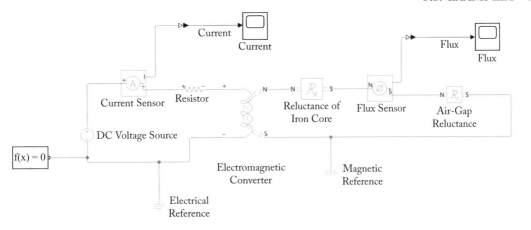

Figure 5.12: An electrically excited magnetic circuit with an airgap.

The electromagnetic converter represents a coil of electric wire. This is the link between an electric circuit and a magnetic circuit. For this model the electrical coil and the electrical circuit that generates the magnetic field are specifically modeled instead of using a source of mmf. Here are the steps to developing the model.

1. Type ssc_new in the Matlab command window to open a new model file.

2. Select and place two Fundamental Reluctances, Electromagnetic Converter, and Magnetic Reference element, Resistor, DC Voltage Source, from the Simscape Magnetics Libraries.

3. Select and place a S_PS and a PS_S element.

4. Select and place the Flux Sensor element and a Current sensor.

5. Set the DC voltage value to 10 V, Resistor value to 1 Ohm, Iron core reluctance to 80, and air-gap reluctance to 39,000 and electromagnetic converter to 500 turns.

6. Connect the elements as shown in Figure 5.12.

7. Leave the Solver setting at "auto" for this simple simulation.

8. Run the simulation for 100 s.

Figure 5.13 shows the flux in the magnetic circuit as it grows from 0 to a steady value. Unlike the previous example the flux does not immediately start with a steady state value but instead grows from zero to the steady state. This is due to the inductance of the coil that is used to generate the mmf. Generating the mmf using the electric coils is a more realistic approach than using a source of mmf like was done in the first example.

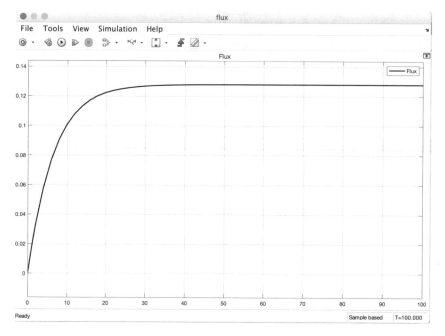

Figure 5.13: The development of flux in the magnetic circuit.

Example 5.3   Magnetic Circuit with Changing Reluctance/Airgap

An interesting phenomenon happens when the airgap or reluctance varies with time. This phenomenon was first observed by Faraday and is formally called Faraday's Law. When the Reluctance changes with time, or the airgap changes with time, thus altering the magnetic flux, a voltage is induced in the electrical circuit that was used to apply the magnetomotive force. And this voltage is in a direction opposite to the applied voltage in the electrical circuit, and is therefore popularly known as the back emf. Faraday's Law is mathematically represented as back emf, e:

$$e = -\frac{d}{dt}(\lambda) = -\frac{d}{dt}(N\phi) = -N\frac{d\phi}{dt}, \tag{5.10}$$

where λ is the flux linkage variable and equates to $N\phi$. The negative sign is indicative of the fact that the back emf acts in the direction opposite to the applied voltage. The concept of back emf and Faraday's Law is a core principle of the workings of many electro-mechanical devices including electric motors. The coil of wire links the electric domain with the magnetic domain. The electric current times number of coils is equal to the magnetomotive force and the rate of change of magnetic flux multiplied by the number of coils is equal to the back emf induced in the electric circuit. So when the magnetic flux changes with time the back emf in the electric circuit results in a changing current in the electric circuit.

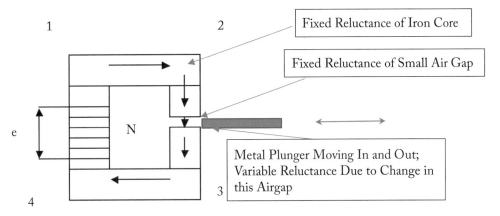

Figure 5.14: Magnetic circuit with a moving plunger that alters the airgap and cuts flux lines.

In this model we will look at a simplified version of this phenomenon. In Chapter 6, we will revisit this in more detail in the context of electric motors as well as a solenoid model.

Here we consider a magnetic circuit that is energized by a coil carrying an electric current. In the airgap in the circuit a magnetic plunger moves in and out to "cut the lines of forces," i.e., alter the magnetic flux. The Reluctance in the circuit can be broken into three parts, the reluctance of the metal core, the reluctance due to the airgap that is in between the plunger and the iron core when the plunger is fully inserted in the gap, and the varying reluctance of a varying airgap as the iron plunger moves in and out and cuts the flux lines. Figure 5.14 shows a schematic of the magnetic circuit. The fixed reluctance of the iron core is assumed to be 80 and the reluctance of the fixed airgap is assumed to be 80,000.

When the plunger, a magnetic material such as magnetic iron, is in the gap the reluctance is very small/negligible; when it is not the empty space is essentially air with a high reluctance. In the following model it is approximated using a variable reluctance block. For this block the airgap varies as per an externally provided signal. The reluctance is accordingly calculated and applied in the circuit. The variable reluctance block parameters are shown in Figure 5.15.

The variable reluctance is given by the equation:

$$\Re(X) = \left(\frac{X(t)}{A_c \mu_0 \mu_r} \right), \tag{5.11}$$

where X is the distance moved by the plunger as a function of time.

In this example $X(t)$ is approximated using a sinusoid that varies between 0 and the full length of the gap assumed to be 0.017 m in this example. This is achieved in the model by summing a sinusoidal function and a step function. The Step function and sine wave inputs are shown in Figure 5.16.

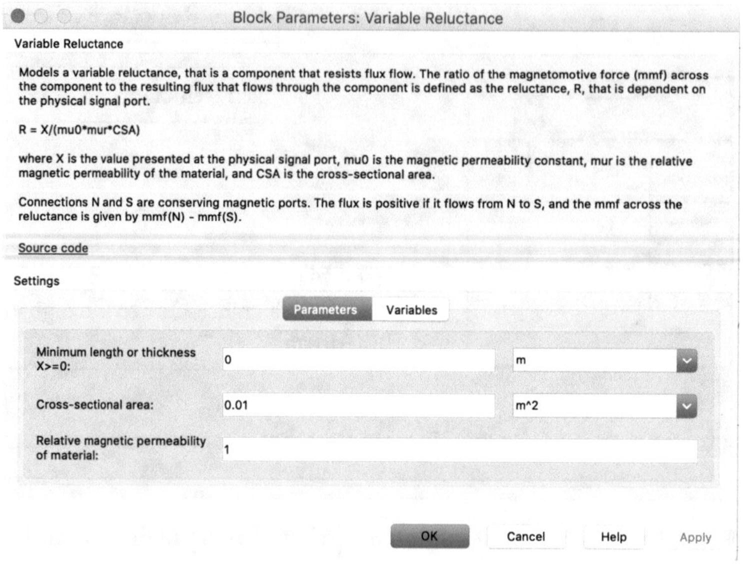

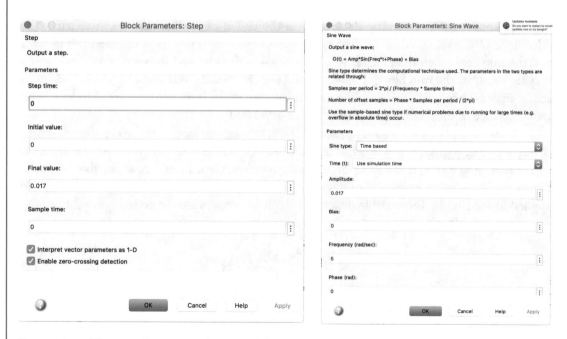

Figure 5.15: Variable reluctance input window.

Figure 5.16: The step function and sinusoid function summed together to form the $X(t)$ for the plunger.

Figure 5.16 shows the input for the variable Reluctance element with the values used in our example. The relative permeability is taken as 1; with the movement of the plunger the airgap increases and decreases sinusoidally. When this variable airgap is zero (plunger is fully inserted) the reluctance is practically equal to zero and when the plunger is completely out the airgap has the maximum length and as a result a high reluctance value.

To put together the model:

1. Type ssc_new in the Matlab command window to open a new model file.

2. Select and place two Fundamental Reluctances, a Variable Reluctance, Electromagnetic Converter, and Magnetic and Electric Reference elements, Resistor, DC Voltage Source, from the Simscape Magnetics and Electric Libraries.

3. Select and place a S_PS and a PS_S element and three Scopes.

4. Select and place the Flux Sensor element and a Current sensor.

5. Select and place a Sine wave and a Step input from the Simulink Library.

6. Set the DC voltage value to 10 V, Resistor value to 1 Ohm, Iron core reluctance to 80, air-gap reluctance to 80,000, and electromagnetic converter to 500 turns.

7. Set the Sine wave parameters and Step input parameters as shown in Figure 5.16.

8. Set the variable reluctance parameters as shown in Figure 5.15.

9. Connect the elements, as shown in Figure 5.17.

10. Use the ODE15s solver in the model configuration parameter menu.

11. Run the simulation for 10 s.

The model is shown in Figure 5.17. Figure 5.18 shows the functional variation of the airgap.

As a result of this varying airgap the flux in the magnetic circuit varies, as shown in Figure 5.19.

As the plunger cuts the lines of forces a back emf is induced in the electric circuit that is proportional to the rate of change of fluxes. This results in the current in the electric circuit to vary with time as well. Figure 5.20 shows the current variation in the electric circuit.

5.4   SUMMARY

In this chapter we discussed some of the basic concepts associated with magnetic circuits and their similarities and differences with electrical circuits. The examples discussed here illustrate the use of magnetic elements to simulate some basic magnetic systems and their behavior.

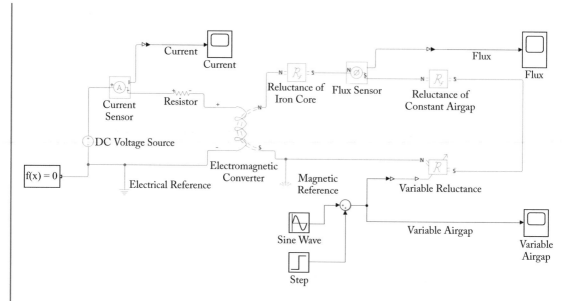

Figure 5.17: **Magnetic circuit model with varying reluctance.**

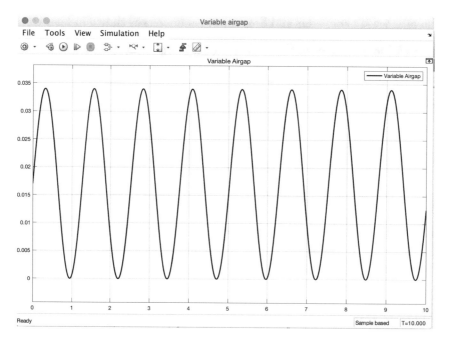

Figure 5.18: **Varying airgap with the movement of the plunger.**

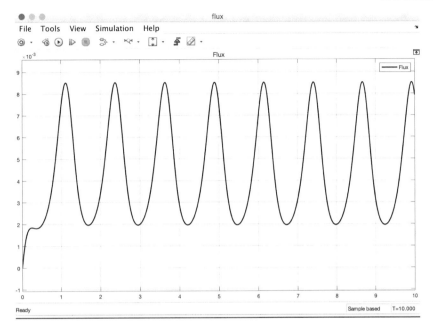

Figure 5.19: Variation of magnetic flux with the variation of the airgap.

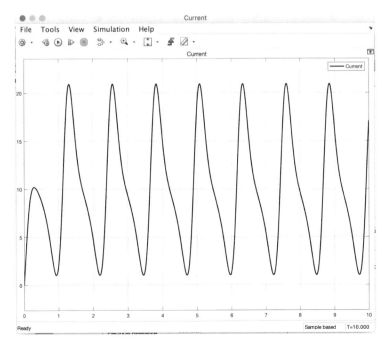

Figure 5.20: Variation of current in the electric circuit with the movement of the plunger.

CHAPTER 6

Modeling Mechatronic Systems (Multi-Domains) and Their Control

6.1 INTRODUCTION

In Chapters 3, 4, and 5 we considered systems which were examples from single engineering domains, such as mechanical, electrical, or magnetic. In this chapter we will build upon the experience from the previous chapters and consider a few examples that are multi-domain, such as mechanical and electrical or mechanical, electrical and magnetic, all within the same system.

The most well-known multi-domain mechatronic system is a DC motor, specifically a permanent magnet DC motor. In this chapter we start with a couple of DC motor models and then consider the control of DC motors for specific tasks and a solenoid valve model that incorporates three domains, electrical, mechanical, and magnetic.

6.2 PERMANENT MAGNET DC MOTOR

Figures 6.1 and 6.2 show schematics for a permanent magnet DC motor. This is the simplest DC motor where the magnetic field is created by a set of permanent magnets that act as the stator and the armature windings are the rotor. Permanent magnet provides a constant magnetic field or flux density. For an armature conductor of length L carrying a current I the force resulting from a magnetic flux density B at right angles to the conductor is BIL (Figure 6.1). With N conductors the force is $F = NBIL$. The forces result in a torque about the coil axis of Fc, if c is the diameter of the armature (or the distance between a pair of opposite forces as shown in Figure 6.1. So the torque may be written as $T = (NBLc)I$. Torque is thus written as $T = K_T I$; I = armature current, K_T is a constant based on motor construction. The armature coil is rotating in a magnetic field, electromagnetic induction will occur and a back emf will be induced. The back emf E is related to the rate at which the flux linked by the coil changes. For a constant magnetic field, this is proportional to the angular velocity of rotation. Hence, back emf is related to flux and angular rotation (in rpm) $E = K_E \omega$; where ω = motor speed in rad/s.

K_T and K_E depend on motor construction and they are of the same magnitude (but of different units). Armature current, at steady state (because the armature inductance behaves like a connecting wire at steady state) $I = (V - E)/R$. R is the armature resistance and E is back

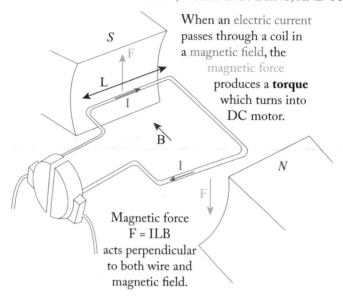

Figure 6.1: A schematic to show the interaction of a current carrying conductor and a magnetic field.

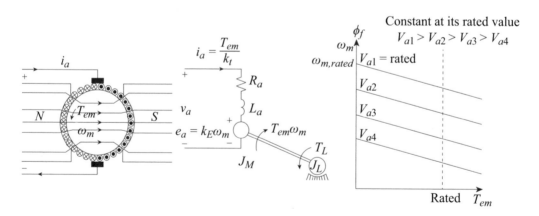

Figure 6.2: Schematic of a motor cross-section, the motor circuit and typical torque vs. speed curves for a PMDC motors.

emf. The Torque, therefore, is $T = T = K_T I = K_T(V - E)/R = K_T(V - K_E\omega)/R$. At start-up, back emf is minimum therefore I is maximum and Torque is maximum. The faster it runs the smaller the current and hence the torque. The motor circuit is shown in Figure 6.2. The current in the circuit is $I = (V - E)/R$ at steady state.

When one considers the unsteady state situation the rate of change of current at the initial state needs to be considered (due to the presence of the inductor). This results in a system of two coupled first order equations (one each on the electrical and mechanical side). All the motor equations together are:

$$\frac{dI}{dt} = \frac{1}{L}(V - E - IR)$$
$$\frac{d\omega}{dt} = \frac{1}{J}(T - T_L - B\omega) \tag{6.1}$$
$$E = K_E\omega$$
$$T = K_T I.$$

The first two equations are the governing differential equations and the next two are the relationships due to the electro-magnetic interaction. The system parameters are: *L (armature inductance)*, *J (rotor inertia)*, *R (armature resistance)*, *B (rotational damping)*, $K_T = K_E$ *(torque or voltage constant)*, and T_L *(load torque)*.

6.3   EXAMPLES

Following are a handful of tips to add new elements into a model.

Tips for adding model elements

1. Use Quick Insert to add the blocks. Click in the diagram and type the name of the block. A list of blocks will appear and you can select the block you want from the list. Alternatively, the Open Simscape Library block can be used to look though the library of all blocks and pick the appropriate one.

2. After the block is entered, a prompt will appear for you to enter the parameter. Enter the variable names as shown in the examples.

3. To rotate a block or flip blocks, right-click on the block and select Flip Block or Rotate block from the Rotate and Flip menu.

4. To show the parameter below the block name, see Set Block Annotation Properties in the Simulink documentation.

Example 6.1   DC Motor Model with Mechanical Load

In this first example we will construct a basic motor model for a PMDC motor. Follow the steps listed next.

1. Type **ssc_new** in the Matlab command window to open a new model file.

2. Select and place a **Current Sensor**, **Ideal Rotational Motion Sensor**, and **Two Torque Sensors**.

3. Using each sensor create a sensing subsystem in the following way: Select a **Out1** block from Simulink library and a **PS_S** converter. Connect the output of the sensor through the PS_S block to the Out1 block. Select all these items, right click and choose to create a subsystem from the menu. Once the subsystem is created, the software automatically adds a **Conn1** and **Conn2**, two connectors for the subsystem to the C and R terminals of the sensor. These connectors will be used to connect this subsystem at appropriate locations in the model. The Torque sensors, and the current sensors are going to measure through variables. The Rotational motion sensor will measure an across variable so use a **Mechanical rotational** reference block to connect to the C terminal for the rotational motion sensor. The R terminal will be connected to the motor output. Figure 6.3 shows these four sensor subsystems.

4. Name these four subsystems, Current measurement, Inertia Torque measurement, Damper Torque Measurement, and Speed Measurement, respectively.

5. Since we will be tracking and plotting a number of variables in the model we would like to minimize the clutter in the model so that it looks clean. The **From** and **Goto** blocks available in the Simulink library are useful blocks to read and write data. These can be used to store data at one location in the model and retrieve the same data at another location in the model. Add four Goto blocks and four From blocks in the model. The output of each sensor will be stored in the Goto blocks and they will be retrieved from the From blocks for plotting. For bookkeeping purposes the Goto and From blocks that correspond to each other should have the same variable name. Double Click on the first Goto block and set the tag to Current and connect it to the current sensor subsystem output. Double click on the first From block and set the tag to Current and connect it to a scope. Name the scope current. Figure 6.4 shows the setup menus for these blocks. Now these Goto-From block pairs will track the current in the circuit and plot it on a scope. Do the same for the other three pairs and name them Inertia Torque, Damper Torque, and Speed, respectively. Connect them to the appropriate sensor subsystem outputs.

6. Put together the model as shown in Figure 6.5. We have set up all the measurement systems.

7. Now we will add all the main elements in the model. Select and place the **Controlled Voltage Source, Inductor, Resistor, Rotational Electromechanical Converter, Inertia, Rotational Damper, Electrical Reference and Mechanical Rotational Reference,** and a **Step Input block**. Connect the elements to create the model shown in Figure 6.6. Notice that the electrical side elements and the model has a different color from the mechanical side.

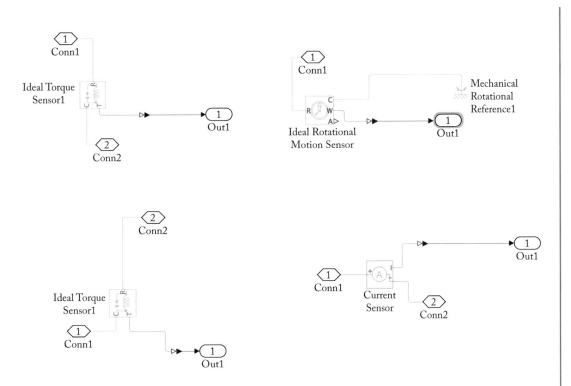

Figure 6.3: Four sensor subsystems.

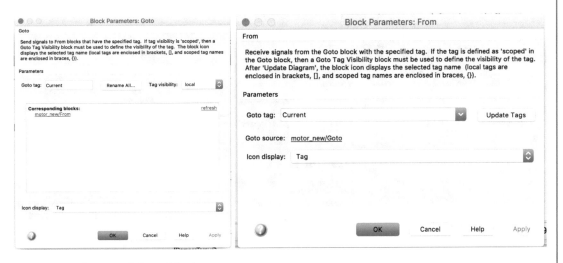

Figure 6.4: Goto and from block setups.

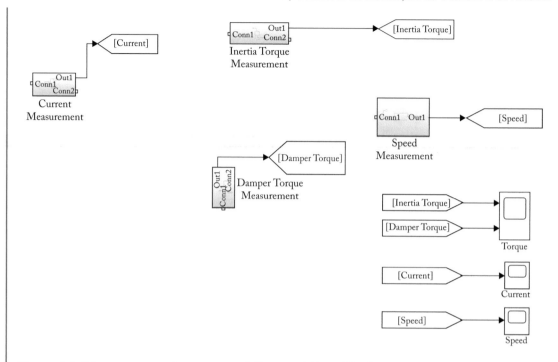

Figure 6.5: Measurement subsystems and Goto and from Block connections.

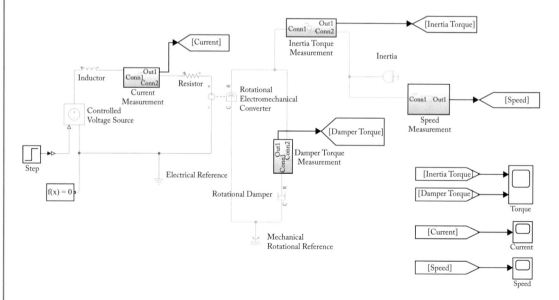

Figure 6.6: Complete motor model.

8. Set the following parameter values for the different elements in the model: Motor Inductance = 1e-6 H, Motor Resistance = 1 Ohm, Rotational Intertia = 0.01 kgm^2, Damping coefficient = 0.001 N/rad/s. Motor constant (torque and back emf) = 0.1 V/rad/s = 0.1 Nm/A (see Figure 6.7). In each case double click on the block and fill out the form to apply these values.

9. Figure 6.8 shows the setting of the step input block. This applies a voltage of 10 V, 1 s after the simulation start time.

10. Use the ODE15s stiff ODE solver by choosing it in the Model Configuration option in the Simulation menu.

11. Run the simulation for 10 s.

Figures 6.9, 6.10, and 6.11 show the simulation outputs, motor current, motor speed, and the torques, respectively. The results clearly show typical motor characteristics. Once the 10 V input is applied the motor starts to speed up from zero speed. At zero speed the back emf is zero and hence the current drawn is the highest. As the motor starts to speed up back emf increases and consequently the motor current drops. With the motor speed reaching a steady value the current settles to a steady value as well. The torque results show similar consistent trends. With the rise in speed the damper torque rises from zero to a steady value. The torque associated with the inertia was highest at the starting when the rotation had to start from zero speed and at steady speed the inertia is rotating at a steady speed and the toque associated with it is therefore zero. The sum of the inertia torque and the damper torque is equal to the total torque output of the motor.

Example 6.2   Motor Model 2
Example 6.1 outlines the development of a motor model that uses basic elements from the library to explicitly model all major components in the motor. Alternatively, Simscape also offers a DC Motor block within its libraries that incorporates most of the electrical as well as mechanical elements of the motor within this one block. In this example the DC Motor block is used to simplify Example 6.1.

So far, all the elements we used in building models were obtained from the foundation library available within Simscape. Beyond the foundation library Simscape has a number of specialized libraries that contain specialized blocks that may be used to model many advanced engineering systems. Figure 6.12 shows us accessing one such library to retrieve the DC motor block. Within Simscape there is a library called Electrical and within Electrical there is a sub directory for Electronics and Mechatronics. The DC-Motor model along with other motor models are available in the Rotational Actuators subdirectory of the Actuators and Drivers directory.

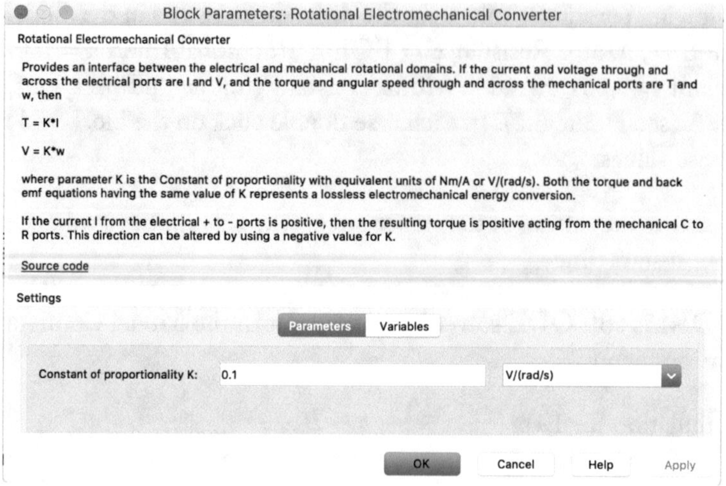

Figure 6.7: Motor torque constant setup.

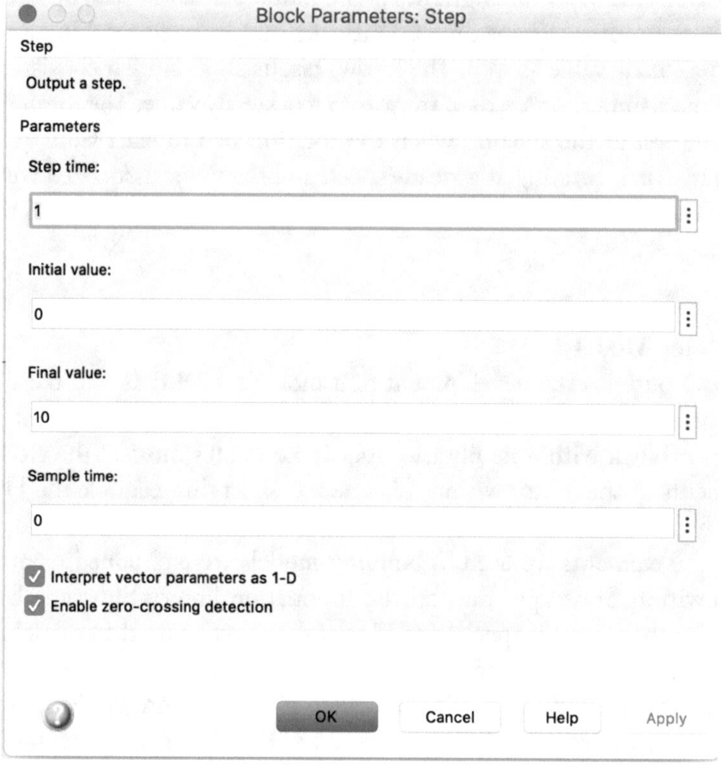

Figure 6.8: Step input used for motor voltage.

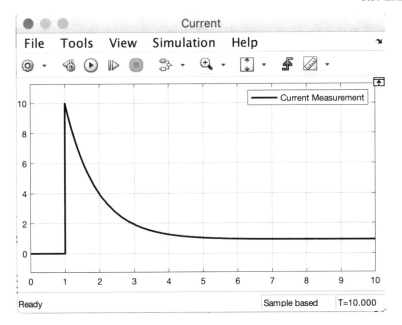

Figure 6.9: Motor current.

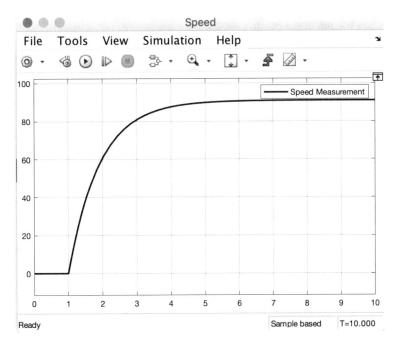

Figure 6.10: Motor speed.

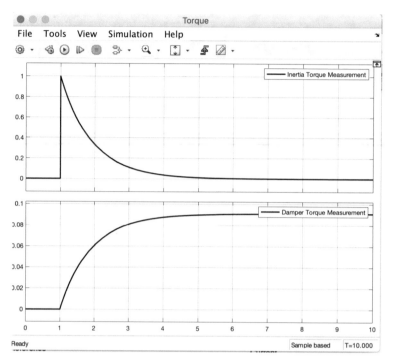

Figure 6.11: Inertia torque and Damper torque.

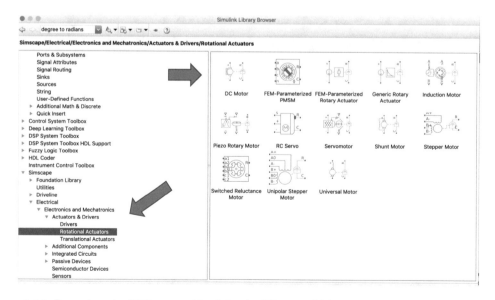

Figure 6.12: Locating the DC motor block in the Electrical library.

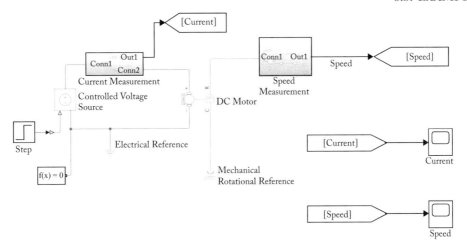

Figure 6.13: Motor model with a DC motor block.

Steps:

1. Open the previous DC Motor model and remove the following items from the model: Resistor, Inductor, Rotational Electromechanical converter, Inertia, and Damper. Also, remove the torque measuring subsystems and the associated scopes and Goto and From blocks.

2. Add the DC Motor block from the Electrical library as shown in Figure 6.12.

3. Connect the blocks as shown in Figure 6.13.

4. Open the DC Motor setup menu and set necessary motor parameters both for Electrical and mechanical components. See Figures 6.14 and 6.15. The paremeters are set to match the values used in the previous example (Motor Inductance = 1e-6 H, Motor Resistance = 1 Ohm, Rotational Intertia = 0.01 kgm^2, Damping coefficient = 0.001 N/rad/s. Motor constant (torque and back emf) = 0.1 V/rad/s = 0.1 Nm/A).

5. The Step input setting of the step input block remains the same as before. This applies a voltage of 10 V, 1 s after the simulation start time.

6. Use the ODE15s stiff ODE solver by choosing it in the Model Configuration option in the Simulation menu.

7. Run the simulation for 10 s.

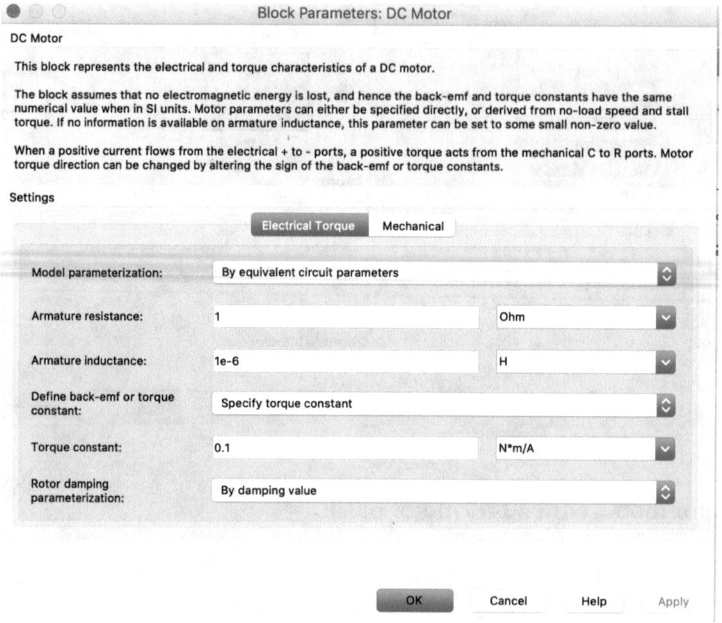

Figure 6.14: Setting the electrical parameters.

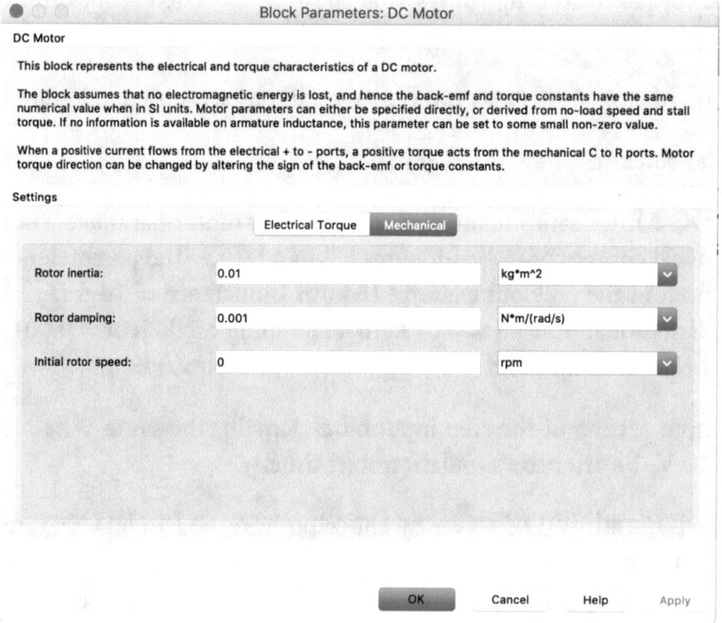

Figure 6.15: Setting the mechanical parameters.

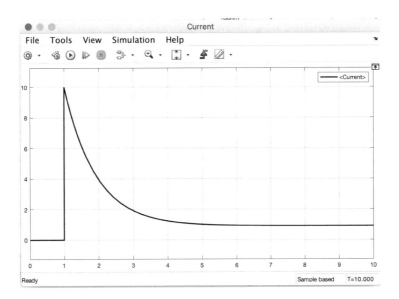

Figure 6.16: Current in the motor.

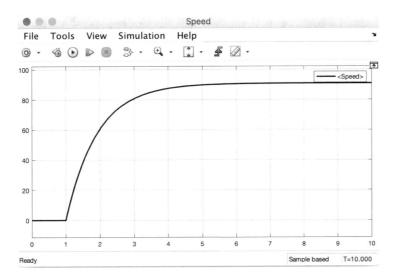

Figure 6.17: Motor speed.

Example 6.3   Control of DC motor for Speed

In this example we will modify the motor model from the first example to add a control loop for speed control tasks. We will use a tunable PID controller provided by Simulink library.

Steps:

1. Open the motor model from Example 6.1 (Figure 6.6).

2. Add a **In1** block and an **Out1** block to the model from Simscape library. Remove the step input in the model and connect the **In1 block** to the controlled voltage source. Connect the **Out1** block to the motor speed output that was connected to the scope.

3. Select everything on the screen and right click to see the menu and choose to build a subsystem. This creates the motor subsystem with one input and one output. Name this subsystem **Motor_Model**.

4. Now add a **Step** Input an **addition-subtraction** block a **scope** and a **PID controller** block from Simulink Library.

5. Connect the elements to create the model shown in Figure 6.18.

6. The Step input provides the desired speed and that subtracts the actual speed to create the error signal. The error signal is passed through the PID controllers to develop the control signal. The control signal is fed into the Motor Model as the voltage input to the model. The actual speed output from the model and the desired speed are plotted together for visualization and judging the effectiveness of the model.

7. Double click on the PID controller to open the block parameter menu. By default it is a PID controller. We can choose to make it a P, PI, or PD controller by picking a desired setting at the top left corner. By default the unit values are assigned to P and I controller parameters and zero for D parameter. Figure 6.19 shows the block parameter menu. We will use these default values to explore the behavior of the controller.

8. Ensure that the ODE15s is the solver chosen in the Configure parameters menu under simulation and run the simulation.

9. Figure 6.20 shows the desired speed, a step change from 0 to 10 at 1 s, and the actual speed. With this initial choice of controller parameters the controller seems to be working pretty well. The actual speed reaches the desired value within half a second.

10. If we want to improve the behavior even more we can use the tune option in the PID block menu (bottom right corner) and tune the PID parameters.

11. Click on the tune button and a plot similar to Figure 6.21 will come up. This provides a preview of the controller performance. There are two sliders at the top of this window. They

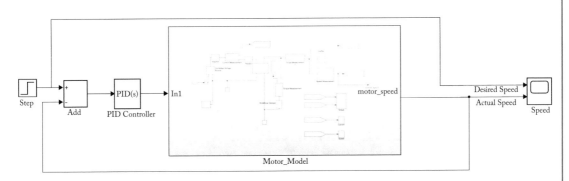

Figure 6.18: Motor speed control model.

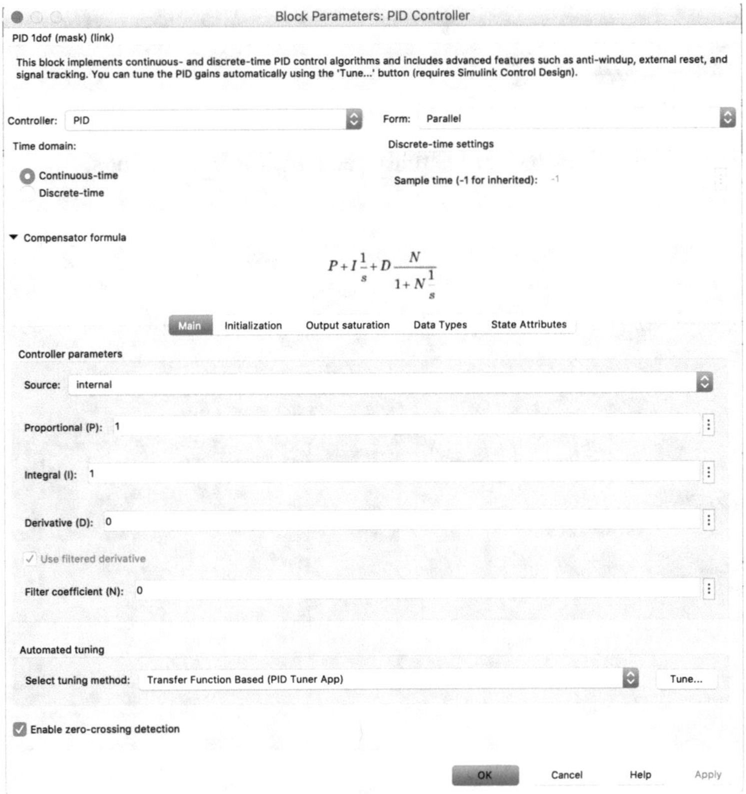

Figure 6.19: PID controller block parameter menu.

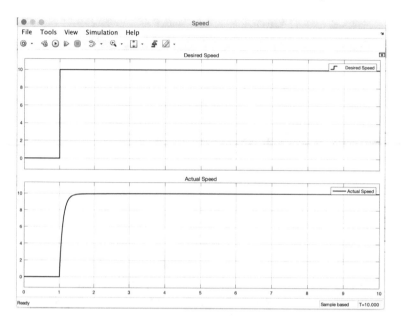

Figure 6.20: Comparison of desired and actual speed with initial settings.

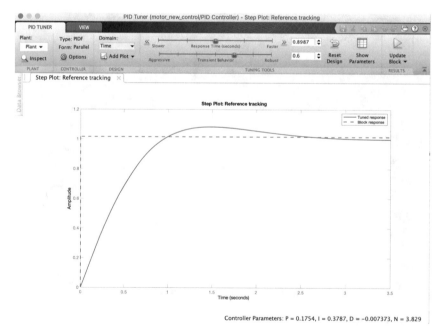

Figure 6.21: PID tuner.

can be adjusted to make the response of the controller faster and more robust. As the slider is moved to the right or left the solid line alters to show how the controller performance changes. When the you are satisfied press the green button to update the parameter values in the PID controller. And rerun the model.

12. Figure 6.22 shows the revised PID settings and Figure 6.23 shows the comparison of the desired and actual speed with this new PID settings and the performance has improved significantly.

Example 6.4   Solenoid with Hard Stop

A solenoid is a very commonly used mechatronic system. It is an actuator of choice where short stroke-length and quick response is needed such as switches and valves. Solenoids are devices that consist of three different interacting domains, electric, magnetic and linear mechanical.

The working principle of a solenoid is as follows: a current passes through a coil that induces a magnetic field around the core of the solenoid. By design air gaps are left in the magnetic path. The plunger which is made of a ferromagnetic material is able to move such that the air gap (and thus the reluctance of the path) is minimized. This movement in the plunger can be used to open or close hydraulic valves used for control applications or perhaps hit a surface to work as a doorbell, etc. The plunger is mounted with a return spring so that when the coils are de-energized the plunger can move back to its initial position thus maintaining the designed gap in the magnetic path. Usually solenoid valves are quite fast acting but the stroke lengths are short. Figure 6.24 shows a schematic of the cross-section of an actual a solenoid. The coils around the inner core, the plunger, and the magnetic path are shown in the picture. Figure 6.25 shows an engineering schematic of the system showing all three domains, magnetic, electrical, and mechanical.

With the passage of the electric current, the energized magnetic field attracts the mass of the plunger which then starts to move. This force of attraction is modeled using the Reluctance force actuator block. Also, when the plunger has moved by a pre-defined amount it hits a hard stop or the back end of the solenoid. In Figure 6.25 schematic this happens when the mass that moves hits the electromagnet and cannot move any more. This is called the hard stop and is also modeled by this same Simscape block. Once the electricity is cut-off the mass will be pulled back by the spring action and will return to its initial position. To begin with (before energizing the solenoid) the plunger is at a position that is not the magnetic equilibrium position. When the solenoid is energized the plunger tends to move to this magnetic equilibrium.

The force expression modeled in Reluctance Force actuator block is derived in the following fashion from the energy relationship.

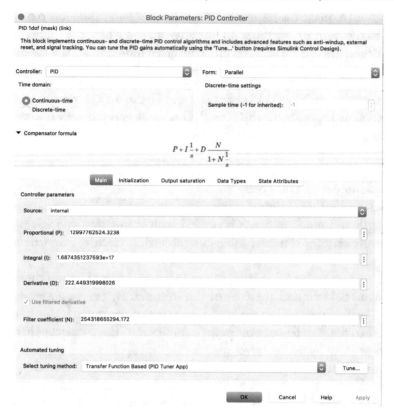

Figure 6.22: Tuned PID parameters.

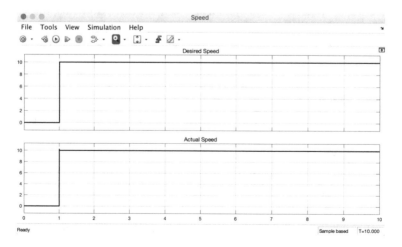

Figure 6.23: Desired and actual speed comparison with tuned controller.

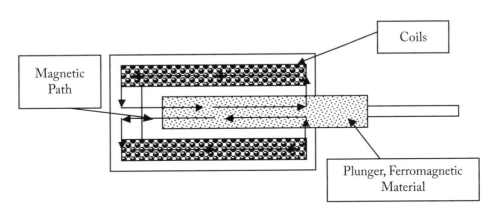

Figure 6.24: Cross-section schematic of a solenoid showing typical construction.

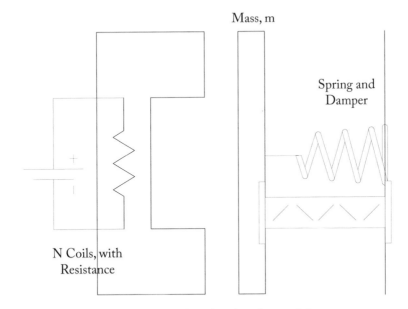

Figure 6.25: Schematic of the solenoid used to develop the model.

Energy stored in a magnetic circuit is

$$Energy = \frac{1}{2}\emptyset^2 \Re, \tag{6.2}$$

where Ø is the flux and $\Re$ is the reluctance.

If we take the derivative of both sides:

$$\frac{d}{dt}(Energy) = Power = \frac{d}{dt}\left(\frac{1}{2}\emptyset^2 \Re\right) \tag{6.3}$$

$$Power = \frac{d}{dt}\left(\frac{1}{2}\emptyset^2 \Re\right) = \frac{\partial}{\partial \emptyset}\left(\frac{1}{2}\emptyset^2 \Re\right)\frac{d\emptyset}{dt} + \frac{\partial}{\partial x}\left(\frac{1}{2}\emptyset^2 \Re\right)\frac{dx}{dt} \tag{6.4}$$

$$Power = (\emptyset\Re)\frac{d\emptyset}{dt} + \left(\frac{1}{2}\emptyset^2\frac{d\Re}{dx}\right)\frac{dx}{dt}. \tag{6.5}$$

In Equation (6.5) the first term on the right side is the magnetic power, product of rate of change of flux, and the mmf. And the second term is the mechanical power, product of velocity (rate of change of x or distance moved by the plunger) and mechanical force. So, $\left(\frac{1}{2}\emptyset^2\frac{d\Re}{dx}\right)$ is the expression for the mechanical force where $(\emptyset^2)$ is the square of magnetic flux and $\left(\frac{d\Re}{dx}\right)$ is the rate of change of reluctance with respect to the gap length x (or movement of the plunger). Figure 6.26 shows the Block parameters for the Reluctance force actuator that is used to model the solenoid. The force expression has a negative sign because the magnetic force is attractive rather than pushing the mass. It also shows the values that are used in this example. Parameters that are needed are: initial gap length; the minimum air gap (usually a very small value that can be considered practically zero); cross-sectional area; relative permeability of material (1 for air); contact stiffness; and contact damping (both these parameters provide an estimate of how elastic the hard stop contact is).

Here are the instructions to put together this model.

1. Type **ssc_new** in the Matlab command window to open a new model file.

2. Select and place a **Current Sensor, Ideal Translational Motion Sensor**, and **Flux sensor** in the workspace.

3. Using each sensor create a sensing subsystem in the following way: Select a **Out1** block from Simulink library and a **PS_S** converter. Connect the output of the sensor through the PS_S block to the Out1 block. Select all these items and right click and choose to create a subsystem from the menu. For the Translational motion sensor add a mechanical reference block and connect to the C-terminal of the sensor block. And for this same sensor attach the Out1 block to the P (displacement) output. Once the subsystem is created, the software automatically adds a **Conn1** and **Conn2**, two connectors for the subsystem to the C and R terminals of each sensor. These connectors will be used to connect each

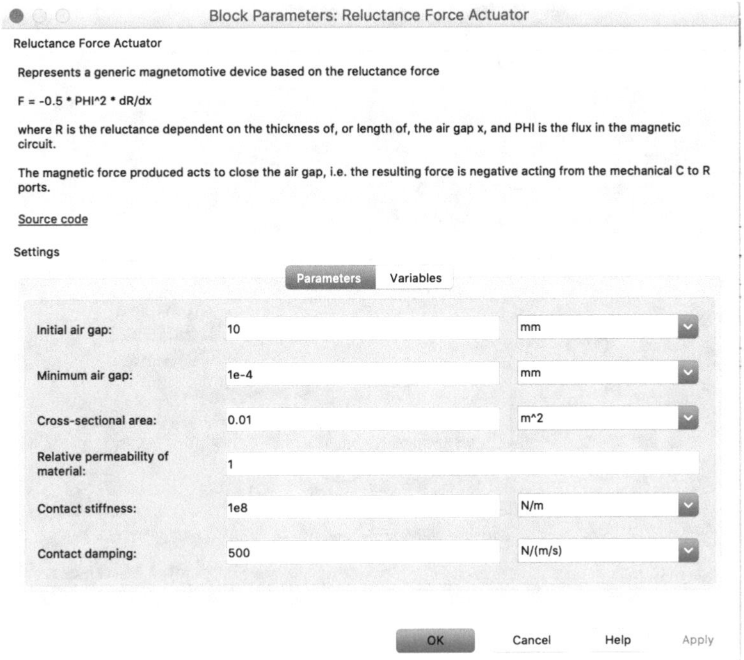

Figure 6.26: Block parameter setup for the Reluctance Force actuator.

subsystem at appropriate locations in the model. The flux sensor and the current sensor are going to measure through variables. The Translational motion sensor will measure an across variable. Figure 6.27 shows these three sensor subsystems.

4. Name these three subsystems, Current measurement, Flux measurement, and Plunger Movement Measurement, respectively.

5. Since we will be tracking and plotting a number of variables in the model we want to minimize the clutter in the model. The **From** and **Goto** blocks available in the Simulink library are useful blocks to read and write data. These can be used to store data at one location in the model and retrieve the same data at another location in the model. Add three Goto blocks and three From blocks in the model. The output of each sensor will be stored in the Goto blocks and they will be retrieved from the From blocks for plotting. For bookkeeping purposes, the Goto and From blocks that correspond to each other should have the same variable name. Double Click on the first Goto block and set the tag to Current and connect it to the current sensor subsystem output. Double click on the first From block and set the tag to Current and connect it a scope. Now this Goto-From block pair track the current in the circuit. Do the same for the other two pairs and name them

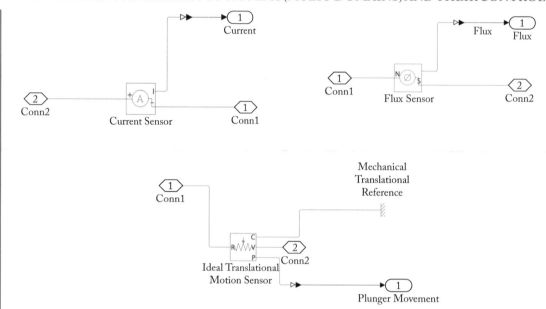

Figure 6.27: Sensor subsystems.

Flux and Plunger Displacement, respectively. Connect them to the appropriate sensor subsystem outputs. Figure 6.28 shows the setup menus for these blocks.

6. Put together the model as shown in Figure 6.29. Now all the measurement systems are setup.

7. Now we will add all the main elements in the model. Select and place the **Controlled Voltage Source, Resistor, Electromagnetic Convertor, Two Reluctances, Reluctance Force Actuator, Mass, Translational Spring, Translational Damper, Electrical Reference and Mechanical Rotational Reference, a Magnetic Reference,** and a **Pulse Generator block.** Connect the elements to create the model shown in Figure 6.30. Notice that the electrical, magnetic, and the mechanical domains have different colors.

8. Set the following parameter values for the different elements in the model: Resistance = 1 Ohm, Reluctance of metal core = 80, Reluctance of Air gap = 8,000, Number of windings in the Electromagnetic converter = 50, Mass = 0.3 kg, Translational Damper = 100 N/m/s, Translational Spring = 3,000 N/m. In each case double click on the block and fill out the form to apply these values.

9. Figure 6.31 shows the setting of the step input block. This applies a voltage of 10 V, 1 s after the simulation start time.

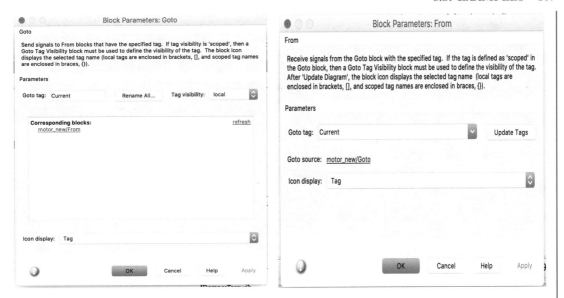

Figure 6.28: Setup menus for From and Goto blocks.

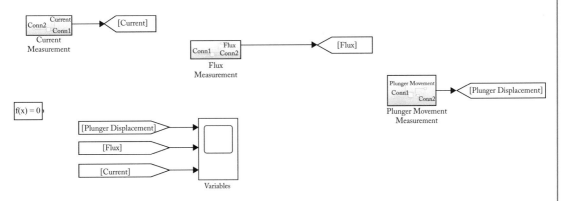

Figure 6.29: Solenoid model with only the measurement subsystems and data handling.

10. Use the ODE15s stiff ODE solver by choosing it in the Model Configuration option in the Simulation menu.

11. Run the simulation for 2 s.

Figure 6.32 shows the system response for this model. The input (pulse generator) applies a 10 V pulse that lasts for 5% of the time period of the pulse (1 s). This current in the electrical circuit and the flux in the magnetic circuit is shown in the output plots. The plunger moves by the whole length of the gap (10 mm) and hits the mechanical stop as shown by the plunger

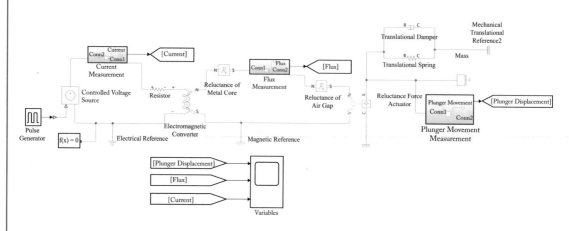

Figure 6.30: Complete Solenoid model.

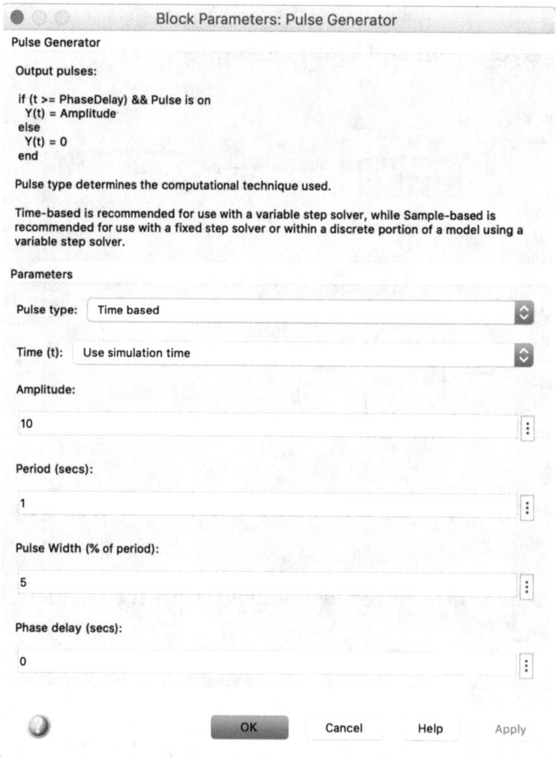

Figure 6.31: Pulse generator setup block.

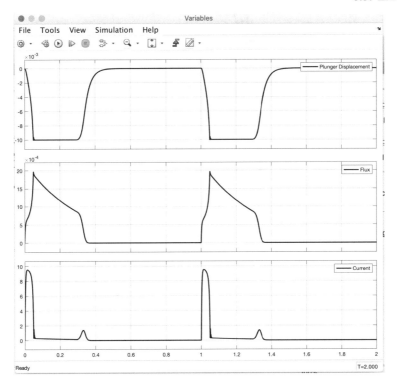

Figure 6.32: System response showing plunger displacement, flux profile and current profiles vs. time.

movement plot. The current does not go immediately to zero after the voltage is turned off due to the inductor effect of the coil. And so the flux takes somewhat longer to die down as does the plunger to move back to its initial position.

Example 6.5   Lifting a Robot Arm and its Control

This last example is a mechatronic system that involves a motor drive that uses a set of gears to move a robot arm in a controlled manner. We will use the previously discussed motor model and modify mostly the mechanical side to simulate the robot arm movement. Figure 6.33 shows a schematic of the system we are modeling. The motor is a Permanent magnet DC motor. The output of the motor is connected to a gear box with a set of gears. The inertia of the two gears are also included in the model. The output of the gear box is connected to a robot arm. The robot arm is a rotating mechanical block whose center of mass is at a distance L from the point of rotation. As a result, this produces a gravitational load moment of $-mgL\sin\theta$ where θ is the angle of rotation of this block/the gearbox output.

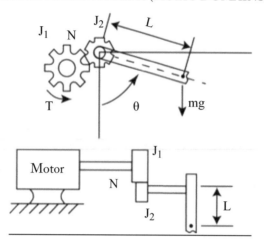

Figure 6.33: Motor drive to control a robot arm through a set of gears.

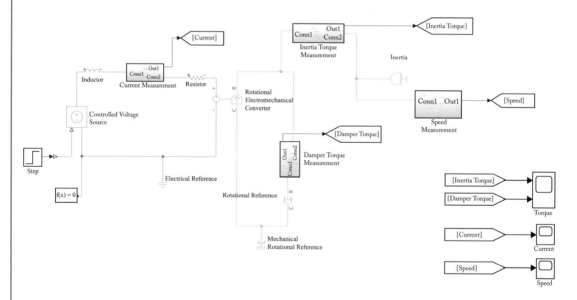

Figure 6.34: Motor model from Example 6.1.

Steps for building the model:

1. Make a copy of the model used in Example 6.1. It is shown in Figure 6.34.

2. Remove the rotational damper in the mechanical side and the torque measurement sensor subsystem associated with it. Remove the Step input block. Add a **Signal builder** block

from Simulink library. Add a **gain** block to the output of the signal builder block and connect it to the input of the Motor voltage source. Set the gain value to 10.

3. Add a second **Inertia** block. Add a **gearbox** block and connect its S terminal to the existing inertia and the O terminal to a second inertia element that is added.

4. The speed measurement block is moved to the O terminal of the gearbox.

5. Modify the speed measurement subsystem by adding an **Out1** block to the angle measurement (there was already a **Out1** block attached to the speed measurement). See Figure 6.35.

6. Connect the model elements as shown in Figure 6.36. **Goto** and **From** Blocks are added to keep track of rotation. This is close to the final stage except for the effect of the rotating inertia.

7. We discussed that the rotating arm will induce a torque of $-mgL\sin\theta$ where θ is the angle of rotation that we are now measuring in the model. Add a **Torque** input Block and a Simulink Function block. Double click the function block and write the expression for $-mgL\sin\theta$. Use $m = 4$ kg, $L = 0.25$ m, $g = 9.81$. The menu is shown in Figure 6.37.

8. Connect the function and torque input block as shown in Figure 6.38. Now the model is completed.

9. Apply the following parameters: Motor Inductance $= 0.002\ H$, Motor Resistance $= 0.5$ Ohm, Inertia $= 0.0851$ kgm^2, Inertia 1 $= 0.37$ kgm^2, Motor Torque Constant $= 0.05$ Nm/A, Gear Ratio $= 2$.

10. Double click on the Signal builder block and create the signal shown in Figure 6.39.

11. Simulate the system for 15 s.

12. Figures 6.40, 6.41, and 6.42 are the torques, speed of rotation of the arm, and the angle of rotation of the arm, respectively. While the torque output matches the voltage input in nature there is significant oscillation observed both for speed of rotation and angle of rotation for the robot arm. After the torques goes down to zero the arm is seen to have steady oscillation of constant amplitude. This is because there is no damping in the system.

13. In a system like this it is often necessary to have a controller to ensure that the movement of the arm is accurate and predictable. So we will add a controller to this model.

14. Add a PID block. We will control the rotation of the robot arm. So use the angle of rotation as the feedback quantity. We will now use the input as desired angle of rotation. Subtract the actual angle of rotation and supply that as the error signal to the PID block. Set the following factors of the Controller. Accept default values for the Proportional, Integral, and Derivative factors. Set the Gain to 1. The model is shown in Figure 6.43.

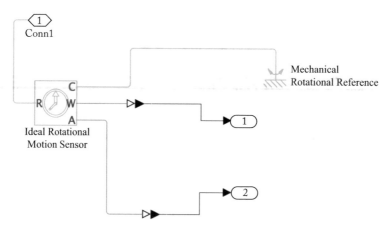

Figure 6.35: Modified speed measurement subsystem.

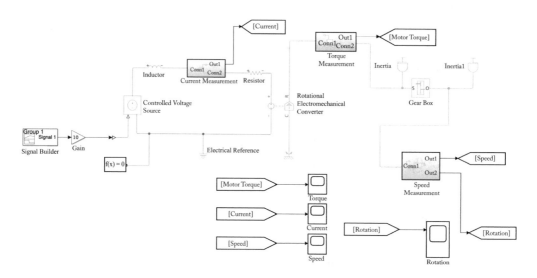

Figure 6.36: Model for motor driving gears (without robot arm).

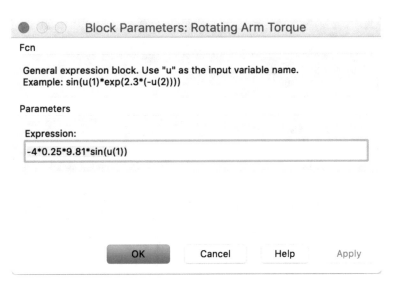

Figure 6.37: Rotating arm torque calculation.

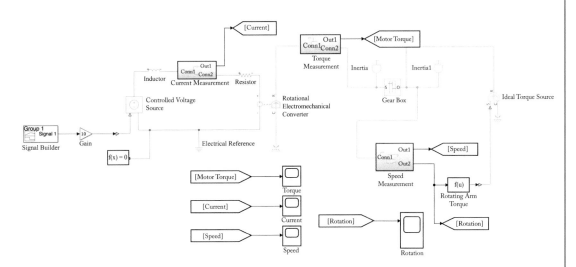

Figure 6.38: Model for motor lifting robot arm.

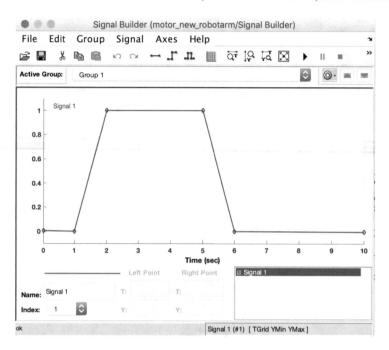

Figure 6.39: Input signal to the motor supply.

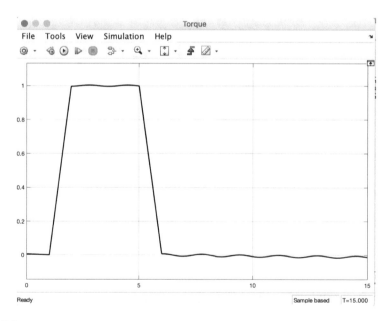

Figure 6.40: Motor torque output.

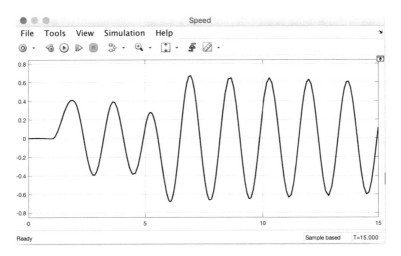

Figure 6.41: Robot arm's speed of rotation.

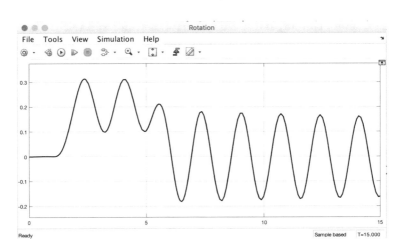

Figure 6.42: Robot arm's angle of rotation.

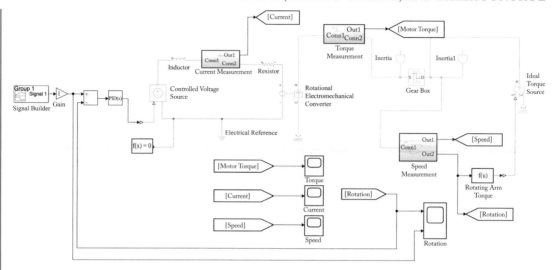

Figure 6.43: Motor and robot arm model with controller.

15. Then run the model with a Stop Time of 15 s. Now Open the PID controller and Tune the controller using the approach described in Example 6.3. Rerun the model.

Figure 6.44 shows a comparison of the actual rotation and the desired rotation with the angle measured in radians. The two plots match very well. Figures 6.45, 6.46, and 6.47 show the Torque output, motor current, and speed of the robot arm, respectively. They all show sharp changes in the profile when the angle of rotation is changing sharply from constant to linear and from linear to constant values. These sudden sharp rise in current would be a matter of concern in any actual system. The controller design needs to take care of this problem using some of the strategies such as cascaded or multivariable control to both control the angle of rotation as well as a reasonable current rise.

6.4   SUMMARY

In this chapter we have looked at a number of examples that are made of subsystems which involve at least two domains. In Simscape, when components from different domains are used the color differences provide a nice visual feedback about the separation of different domains. The examples illustrated how mechatronic systems can be modeled using this tool.

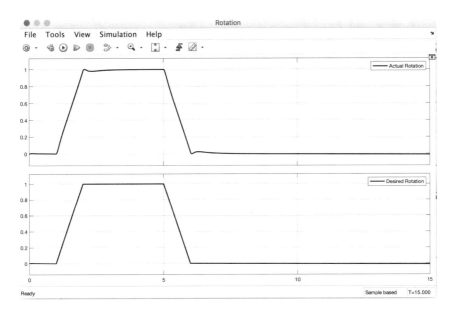

Figure 6.44: Desired and actual rotation of the arm.

Figure 6.45: Motor torque.

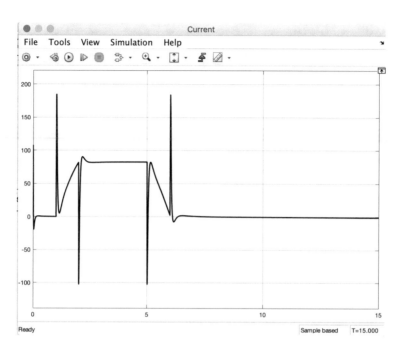

Figure 6.46: Motor current.

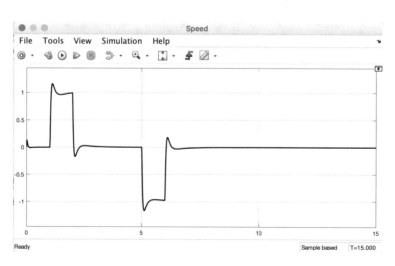

Figure 6.47: Speed of rotation of the robot arm.

CHAPTER 7

Case Studies of Modeling Mechatronic Systems

7.1   INTRODUCTION

In this chapter we discuss a number of models of systems that are somewhat more involved and complex than we have considered so far. Also, these span a broader range of applications ranging from by-wire systems to sub-systems that are relevant to electric vehicles. There is also an example where some rudimentary machine learning concepts are explored. As will be obvious in these examples, we will be using subsystems, models or parts of models that we have used in some of the earlier chapters; for example, the motor model will show up several times in these examples. Therefore the instructions to develop these examples will be a little less prescriptive but no important step will be missed so the reader should be able to develop them from scratch.

7.2   EXAMPLES

Following are a handful of tips to add new elements into a model.

Tips for adding model elements

1. Use Quick Insert to add the blocks. Click in the diagram and type the name of the block. A list of blocks will appear and you can select the block you want from the list. Alternatively, the Open Simscape Library block can be used to look though the library of all blocks and pick the appropriate one.

2. After the block is entered, a prompt will appear for you to enter the parameter. Enter the variable names as shown in the examples.

3. To rotate a block or flip blocks, right-click on the block and select Flip Block or Rotate block from the Rotate and Flip menu.

4. To show the parameter below the block name, see Set Block Annotation Properties in the Simulink documentation.

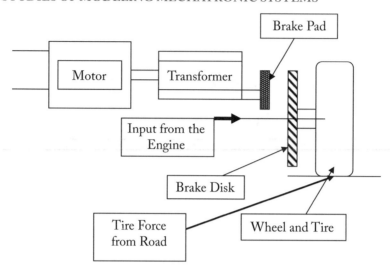

Figure 7.1: Schematic for a brake by wire system.

Example 7.1   Electric Braking System

This example is for an electric braking system. Figure 7.1 shows a schematic of the brake system. The source of the braking effort comes from a motor driven by electric power. The motor is connected to a transformer element which would transform rotational to linear motion. This could be a ball and screw system that is used to apply force on the brake pad. The wheel/tire inertia is modeled as well as a transformer element to transfer rolling velocity to linear velocity at the wheel/ground interface. For wheels that receive direct power from the engine an engine input is included in the model as well. The road resistance needs to be modeled using the constitutive behavior of the tire. These models are relatively complex but can be found in literature and can be incorporated in the model. In this model the resistance is a function of the relative velocity between the longitudinal velocity of the wheel center and the wheel rolling velocity. So a simple rotational damper is used to model this.

To build this model:

1. Type **ssc_new** in the Matlab command window to open a new model file.

2. Build a motor model similar to ones we have used several times in prior examples. Or we can use one of motor models from the prior examples. Use the following parameter values for the motor: $L = 1e\text{-}6\ H$, $R = 0.5$ Ohm, $kt = 150$ V/rad/s, Applied Voltage $= 20$ V in the form of a pulse wave to simulate the braking force (Figure 7.2).

3. Add two **Wheel and Axel** blocks (to be used for Rotation to Translation Transformer or the other way round), **Loaded-Contact Rotational Friction** block, a **torque source,**

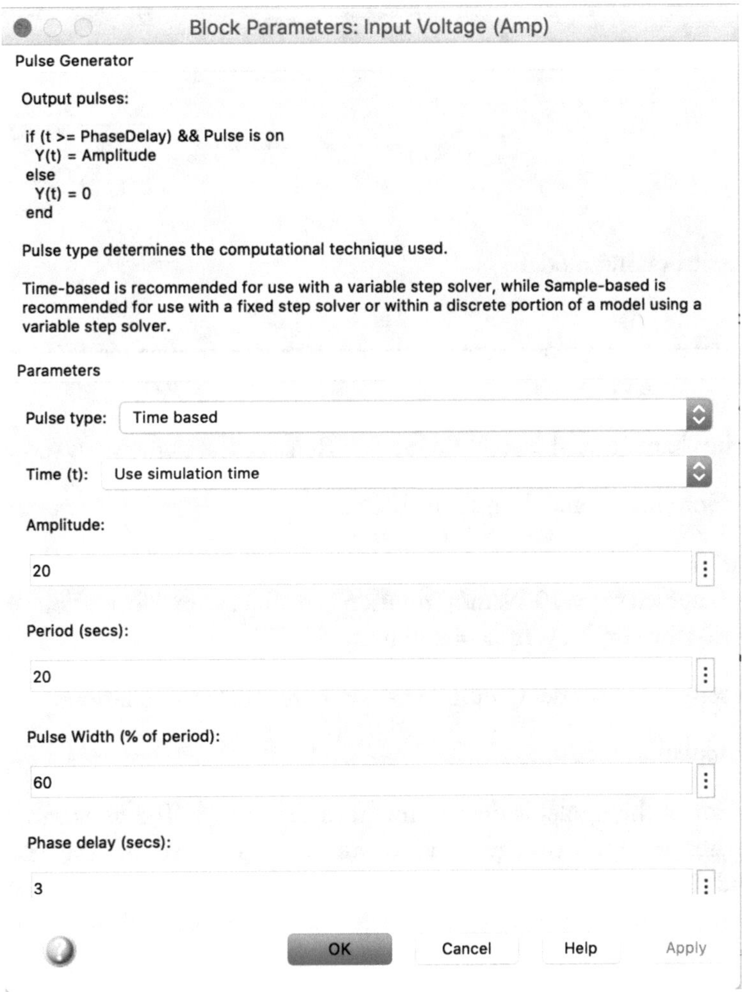

Figure 7.2: Voltage setting for the motor.

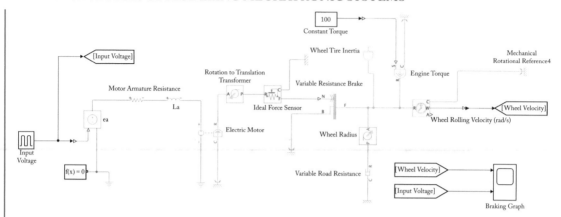

Figure 7.3: Electric braking model.

translational damping block, rotational motion sensor, **Inertia** block, **Ideal Force** sensor, and four reference elements.

4. Connect the elements as shown in Figure 7.3. Rename the elements as shown in the figure.

5. Add Two Goto blocks and Two From Blocks.

6. Wheel Inertia = 10 kgm², Wheel Radius = 0.3 m, Wheel Torque = 100 Nm, Road Damping Coefficient = 10 N/m/s, rotational to linear transformation radius = 0.5 m, Variable resistance brake, torque = 130 mm.

7. Use Auto setting for Model Configuration Parameters for simulation.

8. Run the simulation for 20 s.

The behavior of the model is demonstrated in Figure 7.4. The plots show the rotational velocity of the wheel as motor voltage comes on and is then turned off. The velocity is brought down to zero and then rises as the voltage is removed. The simulation is only a sample representation based on arbitrarily chosen parameters. There was no effort made to check whether these parameters are appropriate for this type of a system. But the model shows how a model for a similar system may be developed.

Example 7.2   Active Suspension
Most vehicle suspensions are passive. This means that the dampers are designed to have a constant damping coefficient. However, we now have technology that can be used to build active or semi-active dampers. These dampers have damping coefficients that can alter adaptively. One well known technology used for active and semi-active dampers is magneto-rheological fluids.

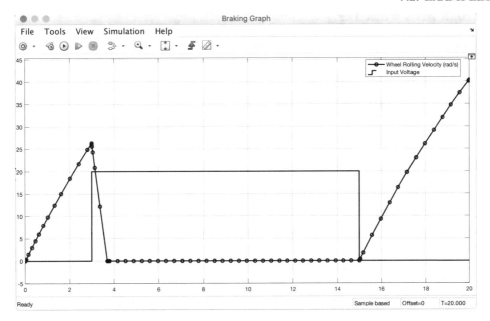

Figure 7.4: Motor voltage profile and the corresponding velocity of the wheel.

In this example, we have not tried to model magneto-rheological dampers but we have made a simpler version by modeling the concept of active and semi-active dampers.

Active and semi-active dampers are designed to address the problems encountered by passive dampers. Many of these active dampers work with the concept called "skyhook damping." Passive damping is based on the relationship that the damping force is proportional to the difference in the velocity of the two endpoints that the damper is attached to (i.e., the relative velocity). In active dampers such as the skyhook dampers the damping force is proportional to the absolute velocity of the mass that is being damped. If B is the damping coefficient:

- Damping force in passive damper $= B *$ (vehicle mass velocity − velocity from road input).

- Damping force in active damper $= -B *$ vehicle mass velocity.

- The force in actively damped system needs to be a negative value so that it works as intended, i.e., as a dissipater of disturbance forces. To implement the active damping in vehicles the absolute velocity of the point where the damper would be attached is used to generate a Source of force and it is fed back at the same location.

In order to demonstrate the effect of skyhook dampers we start with the vehicle suspension model discussed earlier in this text. We use the following parameters in this suspension model:

Vehicle Mass = 2,500 kg

Suspension Mass (Wheel, etc.) = 320 kg

Suspension Spring = 80,000 N/m

Tire Spring = 500,000 N/m

Suspension Damping = 350 Ns/m

Tire Damping = 15,020 Ns/m

The damping coefficiect is purposely kept at a lower value (350 N/m/s). We add a new block element in this model to track the frequency response of this system. A **Bode plot** block is from the Simulink library is added to the model. In the Bode plot setup choose the road disturbance signal as the input and the vehicle mass velocity as the output. The Bode plot is shown in Figure 7.6. The Bode plot provides a measure of the attenuation of the oscillation in the vehicle mass. As the Bode plot shows at lower frequency, 4.5 rad/s, the Bode value is about 25 dB. Since we are plotting 20 log (ratio of output to input), this means that the input is magnified by a factor of $10^{1.25}$. This means that at this frequency the disturbance is significantly magnified when it affects the vehicle body. Figure 7.7 shows the corresponding time series results and shows the velocity and position of the vehicle mass and how it keeps oscillating. At a higher frequency, say 1,000 rad/s the Bode value is about -100 dB. This means that the road disturbance with this frequency content is significantly attenuated in the vehicle mass, or does not affect it. In passive damper design to reduce the peak at the low frequency the damping coefficient is increased. So we arbitrarily increase this to 5,000 and re-run the simulation. Figure 7.8 shows the oscillation of the vehicle mass. It clearly dies down much faster than when the damper was 350 N/m/s. The corresponding Bode plot is shown is Figure 7.9. At the lower fequency, around 4.5 rad/s, the Bode plot peak has gone down significantly. This is expected because of the damping coefficient has been increased. But at the higher frequency, 1,000 rad/s, the Bode value is about -80 dB. This means although the performance at lower frequency has gotten better, the performance at higher frequency has gotten worse. This is the nature of behavior of passive dampers.

The model is now modified by reducing the passive damping back to its original value of 350 and adding a Skyhook damper. The skyhook damper is modeled by adding a force source to the model that applies a force on the vehicle mass. The force is calculated by multiplying the vehicle vertical velocity with a coefficient (-5000), same magnitude as the passive damping used in the previous step. This model is simulated again and the Bode plot is shown in Figure 7.11 and the time response of the vehicle mass is shown in Figure 7.12. The oscillation of the vehicle mass shown in Figure 7.12 which is quite similar to the effect of a passive damper, shown in Figure 7.8. The Bode plot in Figure 7.11 shows that at low frequency, 4.5 rad/s, the reduction from 7.6, the initial bode plot, is similar to what we see in Figure 7.9. But at higher frequency

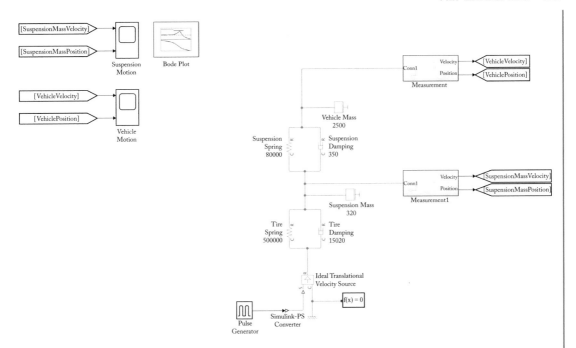

Figure 7.5: Vehicle passive suspension model.

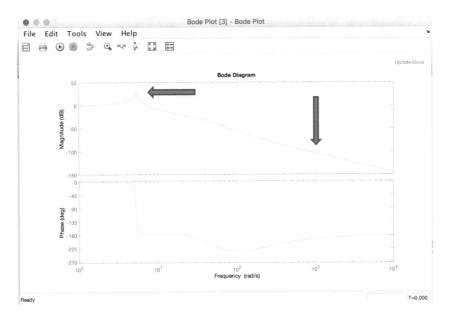

Figure 7.6: Bode plot for a passive damper with damping coefficient of 350 N/m/s.

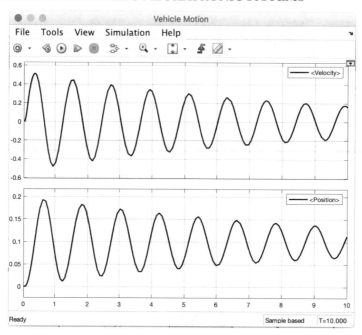

Figure 7.7: Vehicle velocity and position with a passive damper of 350 N/m/s.

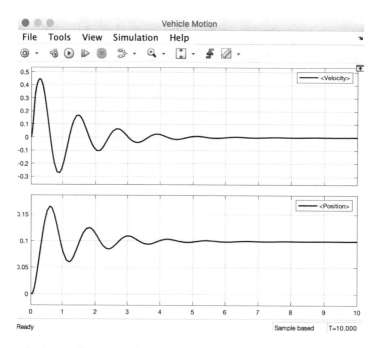

Figure 7.8: The vehicle oscillation with a passive damper of 5,000 N/m/s.

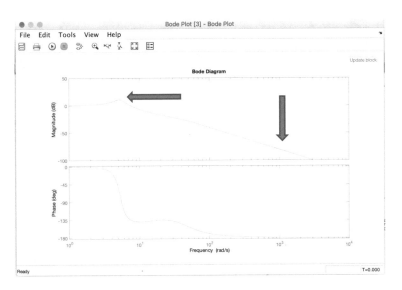

Figure 7.9: Bode plot with passive damper of 5,000 N/m/s.

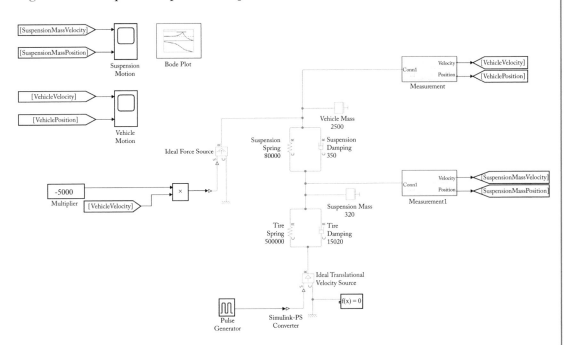

Figure 7.10: Modified model with active damping as in a skyhook damper.

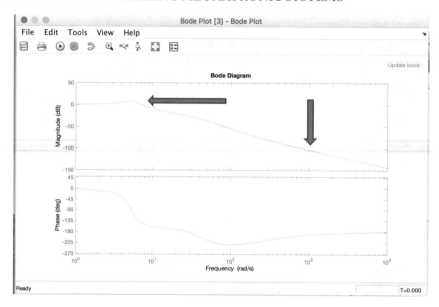

Figure 7.11: Bode plot with skyhook damper of 5,000 N/m/s.

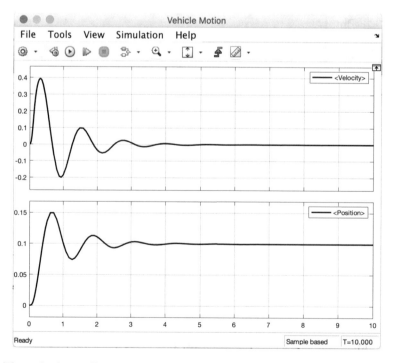

Figure 7.12: The vehicle oscillation with a skyhook damper of 5,000 N/m/s.

with the skyhook damper the Bode value is back to -100 at 1000 rad/s, similar to as seen in Figure 7.6. This means with the skyhook/active damping the improvement of performance seen at lower frequencies does not affect the good performance at the higher frequencies, as was the case for the passive dampers.

Example 7.3 Electric Power-Assisted Steering
    An electric power-assist steering was proposed as a replacement for hydraulic power-assisted steering systems which are in use in most vehicles. This is a simple model of a steering system where the power assist comes from an electric motor rather than a hydraulic pump. The Simscape model is shown in Figure 7.16. There is a driver input in the form of applied angular velocity on the steering wheel. The angle of rotation of the steering column is measured using a rotation sensor and this angle is the input to the assist motor (this is representative of how much of rotation the driver desires). The angular rotation of the steering column is magnified by a a gain factor and the signal is applied as an input to the DC motor. The output torque of the DC motor is added to the input torque from the driver to provide the net output torque that turns the vehicle by steering the wheels. The steering inertia and the steering damping terms in the model are used to model the inertia load and the turning resistance (coming from the contact with the road). Torque sensors are used to track the input torque, assist torque and the total torques.
    To build this model follow these instructions.

1. Type **ssc_new** in the Matlab command window to open a new model file.

2. Add a **Rotation sensor** block, **a gain** block and a Ps_S block and connect them as shown in Figure 7.13 and right click this group and create a subsystem. Name the Subsystem Sensor. Assign a gain value of 5,000 to the gain block.

3. Create a motor model in a manner similar to what was done in the past and add it to the model. Figure 7.14 shows the motor model. Right click the motor model and create a

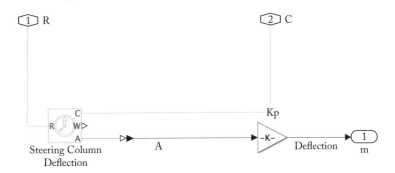

Figure 7.13: Steering sensor sub-model.

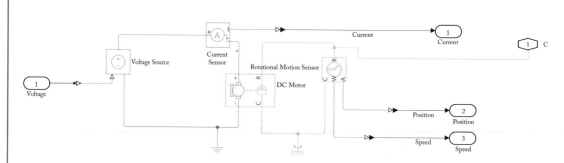

Figure 7.14: Motor sub-model.

Block Parameters: DC Motor

DC Motor

This block represents the electrical and torque characteristics of a DC motor.

The block assumes that no electromagnetic energy is lost, and hence the back-emf and torque constants have the same numerical value when in SI units. Motor parameters can either be specified directly, or derived from no-load speed and stall torque. If no information is available on armature inductance, this parameter can be set to some small non-zero value.

When a positive current flows from the electrical + to - ports, a positive torque acts from the mechanical C to R ports. Motor torque direction can be changed by altering the sign of the back-emf or torque constants.

Settings

| Electrical Torque | Mechanical |

Model parameterization:	By equivalent circuit parameters	
Armature resistance:	2	Ohm
Armature inductance:	2.75e-3	H
Define back-emf or torque constant:	Specify back-emf constant	
Back-emf constant:	0.274	V/(rad/s)
Rotor damping parameterization:	By damping value	

OK   Cancel   Help   Apply

Figure 7.15: DC motor settings.

subsystem. Name it Motor_Subsystem. Set the motor parameters as shown in Figure 7.15. Current, position, and speed are three signal outputs coming out of the model and the motor rotation output is represented by the *C* terminal in the subsystem.

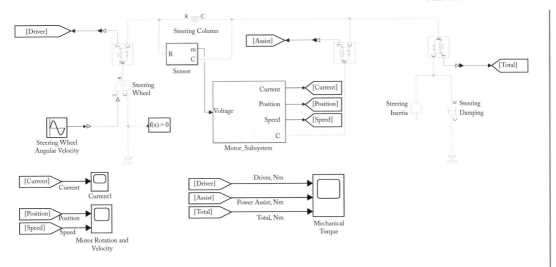

Figure 7.16: Electric power assisted steering model.

4. Now add three **Torque sensors**, one **rotational velocity input**, a **sinusoid function block**, a **rotational spring**, an **inertia block**, and a **rotational damping block** and connect them as shown in Figure 7.16. Rename the blocks as shown in the figure.

5. Assign the following parameter values: Steering Column Stiffness = 100 Nm/rad, Steering Inertia = 1e-5 kgm^2, Steering damping coefficient = 5 Nm/rad/s, Steering wheel angular velocity = A sine wave of amplitude 1 rad/s and frequency of 1 rad/s.

6. Use six Goto blocks and six From Blocks to collect and plot the data as shown in the Figure 7.16.

7. Use Auto setting in Model Configuration Parameters for simulation.

Simulation outputs are shown in Figures 7.17, 7.18, and 7.19. Figure 7.17 shows three torques on a single plot. The assist torque makes up most of the total torque, as needs to be the case. The motor current plot and the motor rotation and speed plots are included here for of the user to get a sense of how they vary at the same time. This example model is to shows conceptually how such a model could be developed. More levels of complexity could be included and more realistic parameter values could be added.

Example 7.4   Simulation and Machine Learning: A Simple Case Study
With the development of advanced algorithms and availability of large quantities of data, applications of Artificial Intelligence (AI) and machine learning is becoming more and more popular in engineering applications. Lately, engineers who specialized in CAE analysis, system

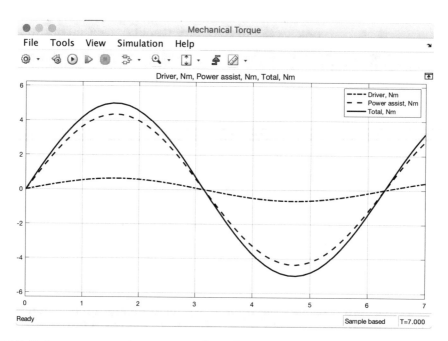

Figure 7.17: Driver torque, assist torque, and total torque.

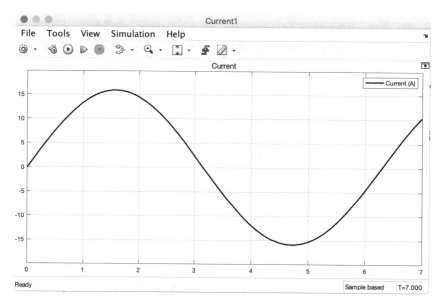

Figure 7.18: Motor current.

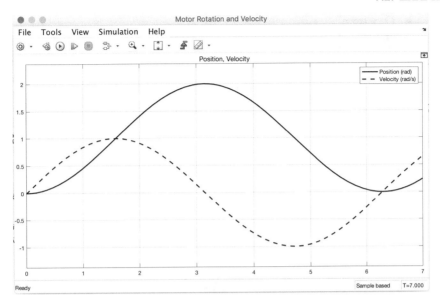

Figure 7.19: Motor rotation and motor velocity.

simulation, and other types of physics-based modeling and simulation are exploring ways to solve problems using various tools of data analytics. One approach that many engineers are taking is to use data from physics-based simulation to develop a machine learning/AI predictive model. Then use this model to make predictions for similar problems in the future in lieu of running simulations.

Figure 7.20 highlights the strategy that is being adopted. With every new problem that a simulation or CAE team encounters they run simulations and use the simulation results to arrive at design decisions. Given that they are perhaps already in possession of large amount of simulation data an alternative approach could be use an appropriate machine learning algorithm to build a model for this type of problem. Provided the model is accurate and robust, when a new problem is encountered in this same arena the machine learning model could be used instead of a full-blown simulation to make design decisions. In Figure 7.20 the solid line depicts current practice. The dot-dash line shows that path of developing a data-fed machine learning model and the dotted line shows how the new model could be used to make design decisions. Here we have picked a simple example to explore this approach.

The example chosen is a Simscape model for a vehicle cruise control device. A fairly simple version of the model is shown Figure 7.21. The Vehicle model consists of a mass and two sources of drag/ friction. A surface contact friction or road resistance and a wind resistance. The road resistance is modeled as a viscous friction that is calculated as vehicle weight * friction coefficient * velocity of the vehicle and is modeled through a friction block. In the model this is depicted

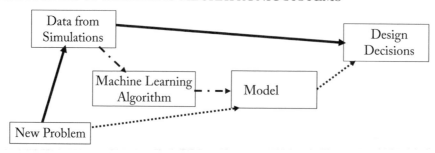

Figure 7.20: Workflow schematic of using simulation data to develop machine learning model.

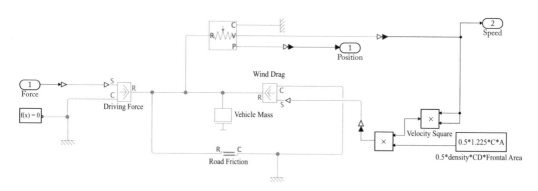

Figure 7.21: Vehicle subsystem.

by the **Road Friction block**. The wind resistance is computed from the wind drag relationship of $\frac{1}{2}\rho C_d A V^2$ where V is the vehicle velocity, A is the frontal area, C_d is the drag coefficient, and ρ is the air density. In this vehicle model the wind drag and the driving force are modeled using two Force blocks. The vehicle mass velocity is obtained through a motion sensor and the velocity value is used in a couple of Simulink mathematics blocks to calculate the wind drag that is fed into the wind drag force block. This subsystem receives signal data for driving force and outputs data about vehicle speed and vehicle position.

The cruise control model uses the vehicle model as a subsystem and links it with a control loop. The control loop uses a PI controller to control the vehicle speed to match a desired speed. The PI controller is optimized for best performance. In the model the Force required to maintain the desired speed is monitored. Figure 7.22 shows the entire model. As switch block and a step function for Force is included in the model for operation without the control loop but for our discussion here the model is always operated with the control loop (i.e., in the mode shown in Figure 7.22)

This simulation study is set up in following fashion: Four input variables are chosen, vehicle mass m, frontal area A, drag coefficient C, and desired speed V. These are varied and the

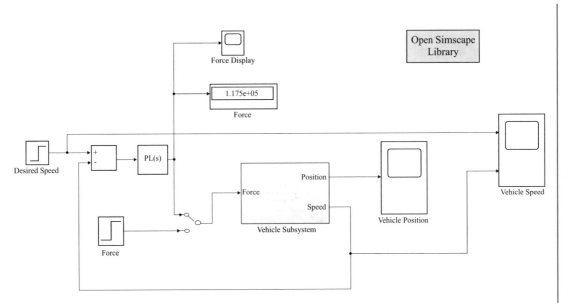

Figure 7.22: Cruise control model.

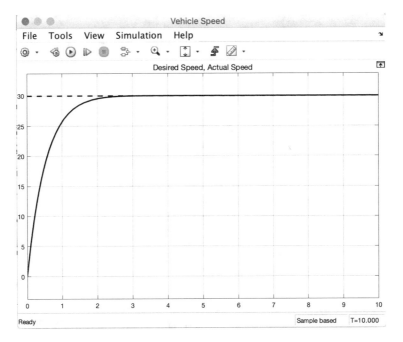

Figure 7.23: Comparison of desired speed and actual speed for one simulation case.

Table 7.1: Simulation settings and driving force (training data)

m (kg)	A (m²)	C	V (m/s)	F (N)
1000	3	0.3	10	1.96E+04
1000	2.5	0.3	10	1.96E+04
1000	2	0.3	10	1.96E+04
1000	3	0.35	10	1.96E+04
1000	2	0.35	10	1.96E+04
1000	2.5	0.35	30	5.84E+04
1000	3	0.35	30	5.83E+04
2000	3	0.3	20	7.83E+04
2000	2.5	0.35	20	7.83E+04
2000	2	0.3	20	7.83E+04
1500	2	0.3	20	5.87E+04
1500	2.5	0.3	20	5.87E+04
1500	3	0.35	20	5.86E+04
1500	2.5	0.3	30	8.79E+04
1500	2	0.3	30	8.80E+04

simulation is run for each setting to determine the force required for each case. Table 7.1 shows the data obtained from simulation.

In this demonstration, we arbitrarily picked a set of representative values for the four parameters and ran the simulation and recorded the force required for the cruise control to behave the same way for all these trial runs. We will call this the training data. The table with shows the parameters and the resultant force for this training data.

Matlab Tool boxes have a series of apps that are used for machine learning model development. These apps are implementation of different algorithms that are used in machine learning applications and range from multi-dimensional regression models to deep learning that involves multi-level Neural Networks. Matlab interface (see Figure 7.24) is setup in a very user-friendly manner where data in the form of tables can be easily used to train various models that are available. Invoking the App shows the many different applications that are available. Under machine learning and deep learning there is an app called Regression learner.

Running the regression learner brings up a screen where the training data can be loaded. The app fits all the different regression models available and as the Figure 7.25 shows, in each case the residual is displayed. The linear regression model has the best fit (lowest residual). Using the "export model option" the model can be exported to the workspace. By default it is called

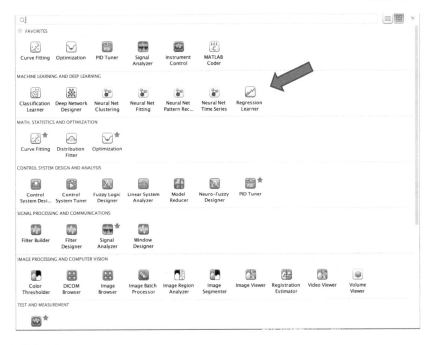

Figure 7.24: Matlab apps menu.

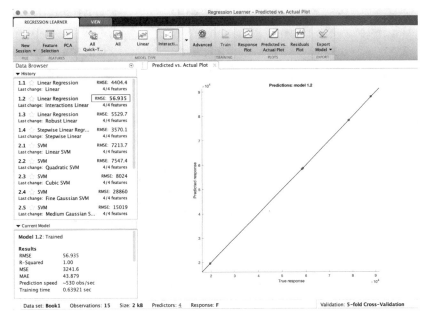

Figure 7.25: Regression models training results.

Table 7.2: Example test data

m	A	C	V
3000	3.1	0.37	20
800	2.2	0.33	35
700	2.7	0.3	21
2700	2.1	0.36	28

Table 7.3: Test data, regression model results, and simulation results

Results				Prediction from Regression Model	Simulation Check	% difference
m	A	C	V	F	F	
1200	3	0.37	20	4.68E+04	4.68E+04	+1.07E-01
1700	2	0.33	25	8.31E+04	8.31E+04	+6.01E-02
1300	2.5	0.3	28	7.11E+04	7.11E+04	-2.81E-02
2200	2	0.36	32	1.37E+05	1.38E+05	+1.67E-01
3000	3.1	0.37	20	1.18E+05	1.17E+05	-1.62E-01
800	2.2	0.33	35	5.45E+04	5.44E+04	-2.76E-01
700	2.7	0.3	21	2.86E+04	2.86E+04	+0.00E+00
2700	2.1	0.36	28	1.48E+05	1.48E+05	+1.49E-01

trainedModel. To test the trained model data can be provided in the form of a table using a table function like the following:

```
>T = table ([3000;800;700;2700],[3.1;2.2;2.7;2.1],[0.37;0.33;0.3;0.36],[20;35;21;28])
>T.Properties.VariableNames = {'m', 'A', 'C', 'V'}
```

The **predictFcn** command used in the following way provides the model output/predicted force from the regression model.

```
yfit = trainedModel.predictFcn(T)
```

Table 7.3 shows the regression model prediction compared with the simulation check. As is easily seen, the regression model predicts the force input extremely accurately. Through this simple exercise we have shown how machine learning can be used in conjunction with simulation to provide an alternative to running simulations for many cases.

7.3   EXAMPLES RELATED TO ELECTRIC AND HYBRID ELECTRIC VEHICLE APPLICATIONS

Applications associated with the design and operation of electric and hybrid electric vehicles are almost all multi-disciplinary in nature. There are five broad topics that are key areas of interest in electrified vehicles: power generation and distribution; energy storage; power electronics and electric drives; control algorithms and controller design; and vehicle dynamics. We are including here two examples that relate to power storage and power electronics.

Battery technology and battery research has become a key focal area in vehicle electrification. It is generally agreed that increasing storage capacity and therefore vehicle range (i.e., distance traveled between charges) is key to getting customers to accept electric vehicles in large numbers. The first example relates to modeling battery behavior. Even though for many modeling exercises we have assumed batteries or sources of power to be providing a constant voltage, real batteries are not quite like that. As more current is drawn from the batteries the state of charge depletes and along with that the potential difference or voltage that the battery can supply. For a realistic system model that has to predict longer term operation, a realistic battery model is essential.

A **Battery** block available in the electric sub-directory of Simscape represents a simple battery model. The block has four modeling variants, accessible by right-clicking the block in the block diagram and then selecting the appropriate option from the context menu, under Simscape > Block choices:

Uninstrumented | No thermal port—Basic model that does not output battery charge level or simulate thermal effects. This modeling variant is the default.

Uninstrumented | Show thermal port—Model with exposed thermal port. This model does not measure internal charge level of the battery but is able to model temperature effect.

Instrumented | No thermal port—Model with exposed charge output port. This model does not simulate thermal effects.

Instrumented | Show thermal port—Model that lets you measure internal charge level of the battery and simulate thermal effects. Both the thermal port and the charge output port are available.

The instrumented versions of the battery model have an extra physical signal port that outputs the internal state of charge. This functionality is used to change load behavior as a function of state of charge, without having to build a separate charge state model.

The thermal port exposes a thermal port, which represents the battery thermal mass. When this option is selected, additional parameters need to define battery behavior at a second temperature.

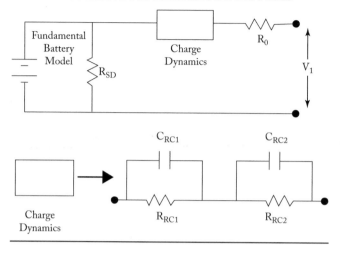

Figure 7.26: Simscape's basic battery model.

The battery model uses a battery equivalent circuit that is made of the fundamental battery model, the self-discharge resistance (as shown in Figure 7.26), the charge dynamics model, and the series resistance R0. Figure 7.26 shows the battery model and the charge dynamics portion explicitly shown. The charge dynamics part uses a series of capacitor-resistor pairs with different time constants to model dynamic behavior.

Example 7.5   Battery Model

The **Battery charge capacity** parameter can be set to infinite when the block models the battery as a series resistor and a constant voltage source. If **Battery charge capacity** is set to Finite for, the block models the battery a charge-dependent voltage source and a series resistor. In the finite case, the voltage is a function of charge and has the following relationship:

$$V = V_0 \left(\frac{SOC}{1 - \beta\,(1 - SOC)} \right), \tag{7.1}$$

where

- SOC (state of charge) is the ratio of current charge to rated battery capacity;

- V_0 is the voltage when the battery is fully charged at no load, as defined by the **Nominal voltage**, parameter; and

- β is a constant that is calculated so that the battery voltage is $V1$ when the charge is $AH1$. The voltage $V1$ and ampere-hour rating $AH1$ are specified using block parameters. $AH1$ is the charge when the no-load (open-circuit) voltage is $V1$, and $V1$ is usually less than the nominal voltage.

Charge Dynamics

You can model battery charge dynamics using the **Charge dynamics** parameter.

- No dynamics—The equivalent circuit contains no parallel RC sections. There is no delay between terminal voltage and internal charging voltage of the battery.

- The Charge dynamics for the battery can be modeled using a series of parallel RC sections. They are denoted in the setup as **First, Second, Third**, etc., **Time constants** depending on how many such RC sections are included in the model. The time constants of each of these sections have to be specified in the block parameters menu.

- Five time-constant dynamics—The equivalent circuit contains five parallel RC sections. Specify the time constants using the **First time constant, Second time constant, Third time constant, Fourth time constant**, and **Fifth time constant** parameters.

- R_0 is the series resistance. This value is the **Internal resistance** parameter.

- R_{RC1} and R_{RC2}, etc., are the parallel RC resistances. Specify these values with the **First polarization resistance** and **Second polarization resistance** parameters, respectively.

- C_{RC1} and C_{RC2} are the parallel RC capacitances. The time constant τ for each parallel section relates the R and C values using the relationship $C = \tau/R$. Specify τ for each section using the **First time constant** and **Second time constant** parameters, respectively.

Modeling Thermal Effects

For thermal variants of the block, you provide additional parameters to define battery behavior at a second temperature. The extended equations for the voltage when the thermal port is exposed are:

$$V = V_{0T} \left(\frac{SOC}{1 - \beta\,(1 - SOC)} \right) \tag{7.2}$$

$$V_{0T} = V_0\,(1 + \lambda\,(T - T_1)), \tag{7.3}$$

where:

- T is the battery temperature;

- T_1 is the nominal measurement temperature;

- λ_V is the parameter temperature dependence coefficient for V_0; and

- β is calculated in the same way as described before, using the temperature-modified nominal voltage V_{0T}.

The internal series resistance, self-discharge resistance, and any charge-dynamic resistances are also functions of temperature:

$$R_T = R\left(1 + \lambda_R\left(T - T_1\right)\right),\tag{7.4}$$

where λ_R is the parameter temperature dependence coefficient.

All the temperature-dependent coefficients are determined from the corresponding values you provide at the nominal and second measurement temperatures. If you include charge dynamics in the model, the time constants vary with temperature in the same way.

The battery temperature is determined from a simple summation of all the Ohmic losses included in the model:

$$M_{th}\dot{T} = \sum_i \frac{V_{T,i}^2}{R_{T,i}},\tag{7.5}$$

where:

- M_{th} is the battery thermal mass (mass * specific heat capacity);

- i corresponds to the ith Ohmic loss contributor. Depending on how the model is configured the losses include: Series resistance, Self-discharge resistance, First charge dynamics segment, Second charge dynamics segment, Third charge dynamics segment, Fourth charge dynamics segment, Fifth charge dynamics segment;

- $V_{T,i}$ is the voltage drop across resistor i; and

- $R_{T,i}$ is resistor i.

Follow the following steps to develop the battery model (this example was developed using an example in the Simscape help manuals).

1. Type **ssc_new** in the Matlab command window to open a new model file.

2. Select and place a **Constant Block**, a temperature source block, a **temperature measurement sensor**, and a **Convective Heat transfer** block from the Heat transfer sub folder in Simscape. Add a Conn1 block and a Thermal reference block. Connect them as shown in Figure 7.27. The battery is the source of heat in this model and the heat generated will be transferred by convection to the surroundings. Set the surrounding temperature to be 25°C. In the convective Heat transfer block set the area to be 0.1 m^2 and the convective heat transfer coefficient to be 0.1 W/m^{2}K. Now select all the items in this thermal model and make a subsystem by right clicking and choosing from the menu. Call this subsystem **Thermal Aspect**.

3. Choose a **Battery block** from the Electrical folder in Simscape under Sources. Right click on the block and choose Simscape => Block Choices => Instrumented | Show thermal port.

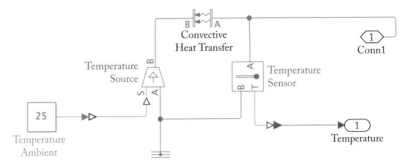

Figure 7.27: Thermal subsystem.

4. The Battery block settings is a complex set of tables because of the many options that are available. For this model the choices made are shown in Figure 7.28. Each pane is for a particular part of the setting.

5. Add a **Voltage Sensor**, a **Current sensor**, and a **Current source**. Instead of having a load circuit in the model we will use a current source to drive a current load in the circuit to explore its effect on the battery.

6. Add a **Signal generator** and rename it Load Current Waveform. Figure 7.29 shows the waveform that is used. This waveform is similar to a waveform that might be see in the current drawn in an electric vehicle.

7. Add four **From** block and four **Goto** blocks to the model. Create four pairs of From-Goto blocks by naming them Temperature, Voltage, Current, and SOC.

8. Attach all the elements as shown in Figure 7.30.

9. The battery has a terminal (q) which provides data on the amount of charge left in the battery. The SOC (state of charge) is calculated using this data and with a multiplier/gain block. The gain factor is (1/(100*3,600)). The 3,600 is to convert hours to seconds and 100 is the capacity of battery in hrA.

10. Set the solver setting to Auto and run the simulation for 3,500 s.

Figures 7.31, 7.32, 7.33, and 7.34 show the Current Temperature, SOC, and the Voltage across the battery respectively, for this model. The SOC plot show how the state of charge drops from initial value of 0.8 to 0.55 over this length of time. The temperature starts at 25°C and goes up to 26.2°C.

Example 7.6   Power Electronics
Power Electronics is a key technical area that plays a very important role in the operation of electric and hybrid electric vehicles. Some of the core tasks of power electronics involves

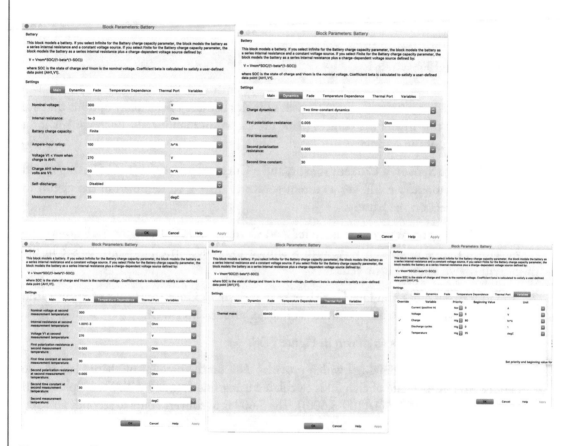

Figure 7.28: **Battery model setup.**

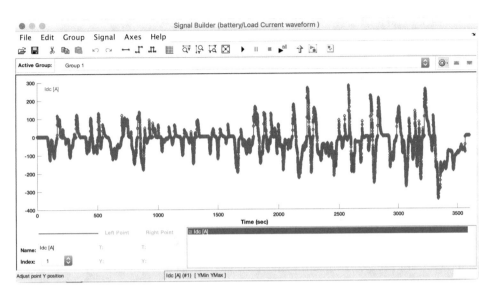

Figure 7.29: Current profile.

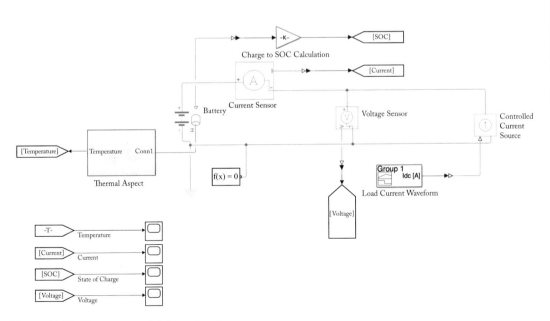

Figure 7.30: Battery model in a circuit.

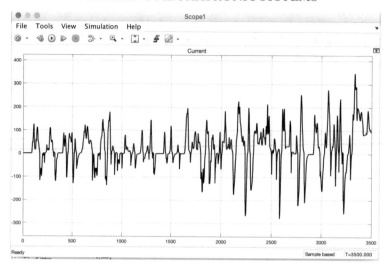

Figure 7.31: Current profile or current load in the circuit.

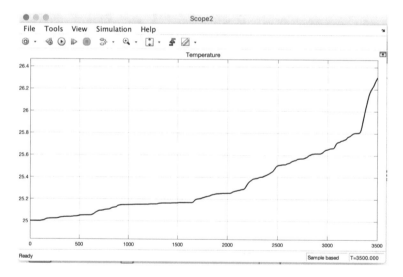

Figure 7.32: Temperature in the battery.

boosting voltages, both DC and AC, i.e., raising the available battery voltage to a higher value usable by the electric drives; or converting DC input to 3-phase AC because the battery is a source of direct current whereas the traction motors run on 3-phase alternating current. Two simple examples are included here; one on DC to DC conversion of a lower voltage to a higher DC voltage through a boost converter and a second one on converting DC supply to 3-phase AC.

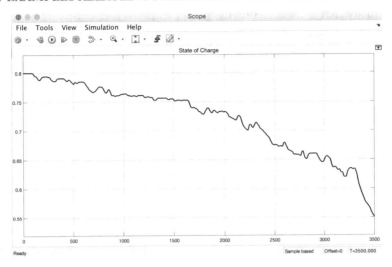

Figure 7.33: State of charge in the battery.

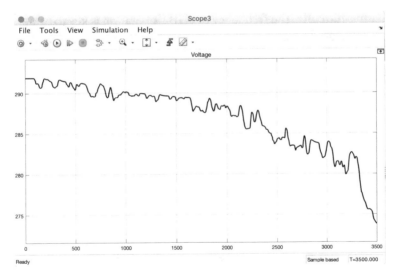

Figure 7.34: Voltage output from the battery.

BOOST CONVERTOR

Power electronic circuits are heavily dependent on electronic/semiconductor switches which enable circuit designers to perform fairly complex tasks such as boosting or reducing DC voltages or transforming DC supplies into single-phase or 3-phase AC supplies. In EV and HEV applications the source of power is a battery which supplies DC voltage whereas the motor and the drive technology that is used to provide power to the vehicles for vehicle motion runs on 3-phase AC. Power electronic circuits continuously has to switch between AC and DC and therefore these circuits are vital parts of the HEV and EV power trains. Two of these electronic switches that are very commonly used in power electronic circuits are MOSFETs (Metal Oxide Semiconductor Field Effect Transistor) and IGBTs (Insulated Gate Bi-polar Transistor). There are many references in open literature about design, construction, and function of these components. We will not get into that discussion here. We will use them to construct models that will demonstrate a particular functionality. We have used the IGBT block available within the electrical subdirectory of Simscape for both examples. In MOSFET and IGBT applications these electronic switches are used by turning them on and off at different frequencies. Based on how they are arranged in a circuit, this allows or stops current along different paths or accumulates energy in storage devices such as an inductor or a capacitor. The specific circuit design then allows the system to provide the desired functionality.

Figure 7.35 shows the model for a Boost converter. The goal of this power electronic circuit is to boost a DC voltage. The supply voltage is from the constant voltage source in the circuit and the output is across the load resistor where a voltage measuring sensor is recording the voltage. The particular functionality of this circuit is achieved by turning the IGBT on and off to alter the current path intermittently. The frequency of IGBT operation is key to achieving the desired boost. All the elements used here are available in the Simscape library either in the fundamental directory or the electrical directory. The parameters used are:

$L1 = 400E - 6$, $R1 = 0.001$ Ohms, DC Voltage $= 100$ V, $C = 25$ E-6 F, Load Resistor $= 60$ Ohms.

Figure 7.36 shows the setup menu for the Pulse generator. This shows that the pulse has a period of 100 micro sec, i.e., the switch on-off frequency is 1/(100E-6) $= 10$ KHz. Also the switch is on for 50% of the cycle time and off for 50% of the cycle time. This on-off time period

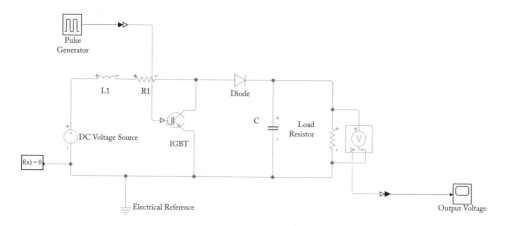

Figure 7.35: Boost converter circuit model.

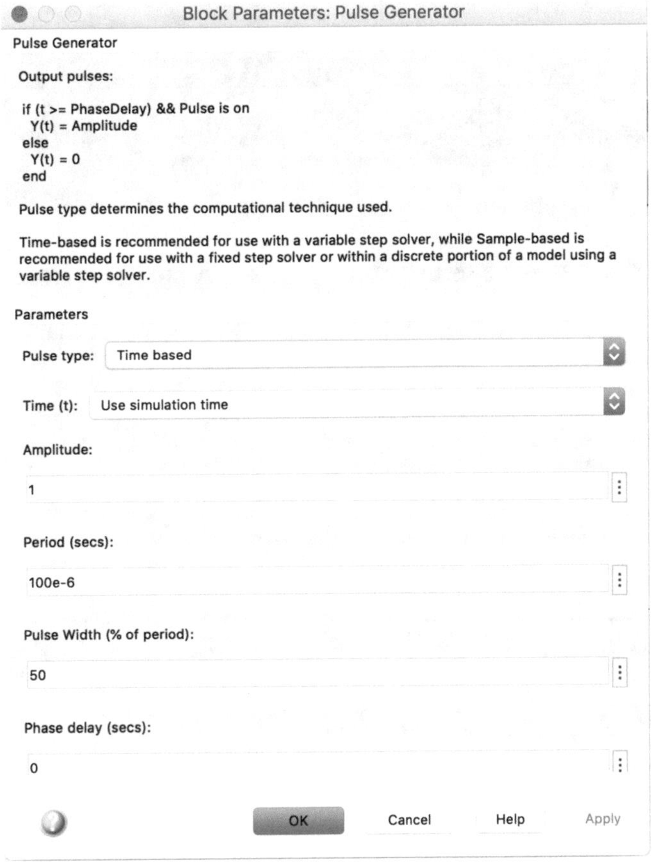

Figure 7.36: Pulse generator setup.

is an important parameter in the IGBT behavior and outcome. The boost voltage is given by:

$$V_{boost} = V_{in}\left(\frac{1}{1-r}\right),$$ (7.6)

where r is the fraction of time the switch is on and in this example, it is 0.5. Thus, the output voltage across the Load resistor would be 200 V in this example. Figure 7.37 shows the output voltage to be about 200 V.

INVERTER

Figure 7.38 shows the inverter circuit that uses six IGBTs to convert a DC input voltage to a 3-phase AC output voltage. The figure shows the circuit which can be found in many references. The load consists of three resistors shown in the model as resistor A, B, and C. For a balanced

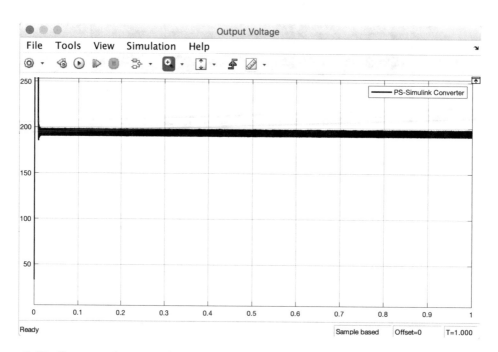

Figure 7.37: Output voltage in a boost converter.

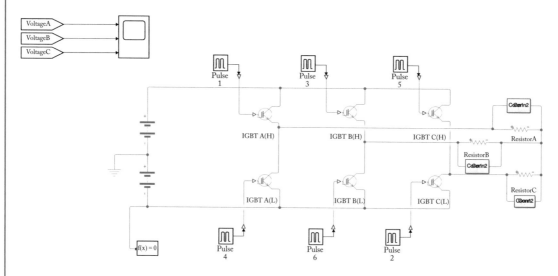

Figure 7.38: Inverter model.

3-phase circuit these resistances are the same. In typical applications the load is usually a resistor-inductor combination (coils of motors, etc.). But to keep things simple we are using only resistive loads for this example. In the model there are three subsystems connected in parallel to load resistors. These are measurement sensor subsystems for measuring voltage across each of these resistances. The content of one of the subsystems is shown in Figure 7.39. Like in many of the past models discussed in this text, Goto and From blocks are being used to collect and plot data (the 3-phase voltage). The supply voltage is 200 V per battery, i.e., 400 V. Each load resistor is 1 ohm. Figure 7.40 shows the block parameters of each IGBT electronic element. Once again, the switching frequencies of the six IGBTs and their phase differences make the circuit work in the desired fashion. Figure 7.41 shows the setup for pulse 1 which is applied to IGBT A. The pulse amplitude is 4 (this should be nominally greater than the threshold voltage of the IGBT (which is set at 0.5) in the previous figure. The period is set at 0.02 s. This means the frequency is $1/0.02 = 50$ Hz. The pulse is on for 50% of time and turned off 50% of the time. The six IGBTs are setup the same way but set apart by a phase difference. Each one is 60° apart in the following fashion:

Phase of Pulse1 = 0*0.02/360

Phase of Pulse2 = 60*0.02/360

Phase of Pulse3 = 120*0.02/360

Phase of Pulse4 = 180*0.02/360

Phase of Pulse5 = 240*0.02/360

Phase of Pulse6 = 300*0.02/360

The simulation is run for 2 s and the output voltages are shown in Figure 7.42. The goal of this simulation was to demonstrate how an inverter circuit converts a single phase DC input to a 3-phase AC output. Figure 7.42 clearly shows that.

7.4   SUMMARY

In this chapter we considered a number of modeling cases studies. The systems considered ranged from power electronic devices to by-wire systems. All the simulation exercises illustrate the power of Simscape as a tool to model mechatronic systems. An example where simulation is combined with regression analysis to develop a machine learning model of an engineering system with several variables was also included. This example illustrates how existing or new physics-based simulation results can be used to generate data that can train a machine-learning model. Such a machine-learning model can then be used to do quick analysis specially at the early stage of a system design process to make decisions. The examples covered in this book illustrate how

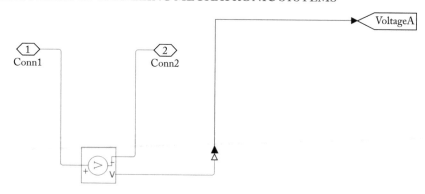

Figure 7.39: Voltage measurement sub-model.

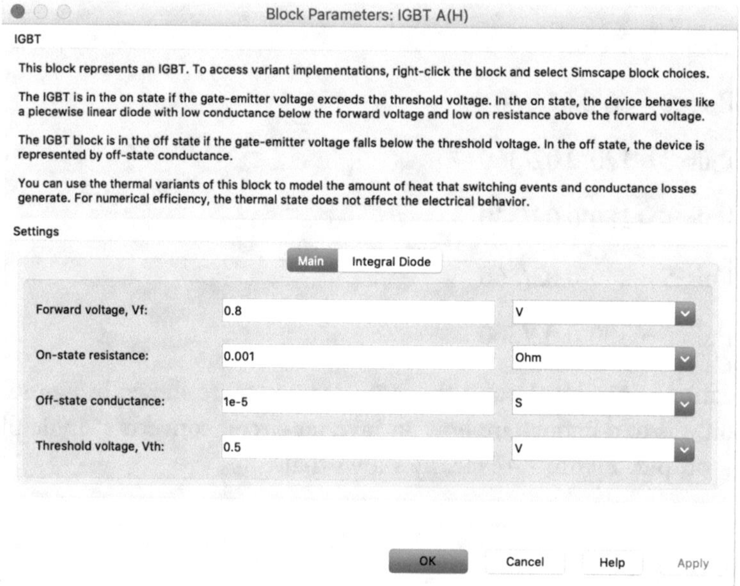

Figure 7.40: IGBT parameters.

a variety of problems can be solved using Simscape. Physics-based models can provide a lot of insight into system dynamics for the designer before even the device is built.

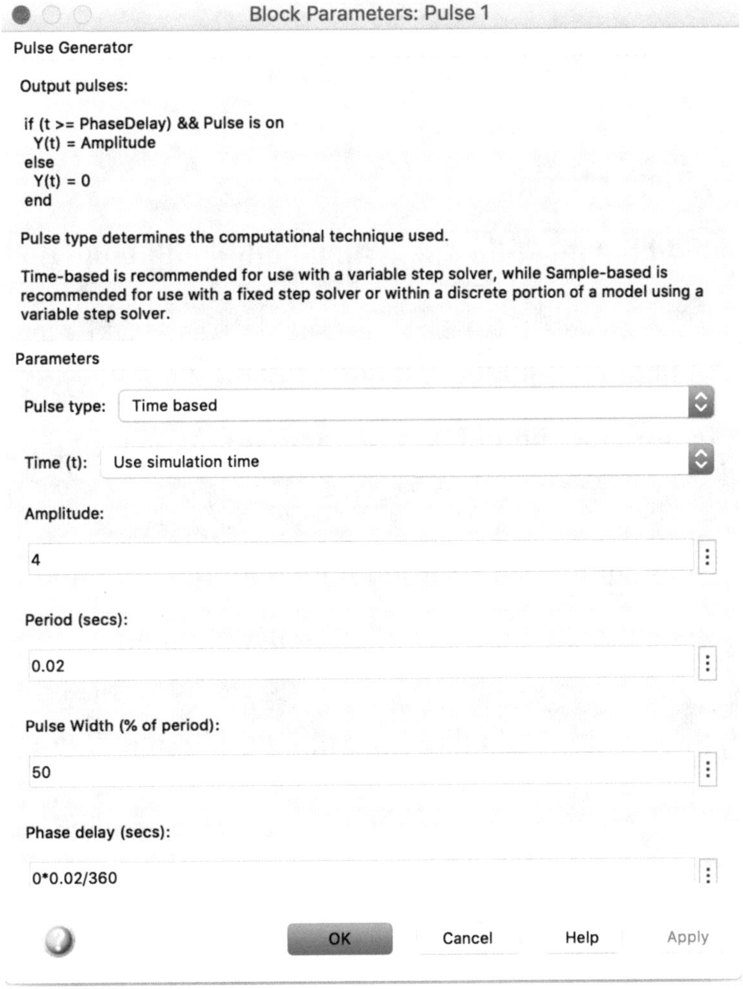

Block Parameters: Pulse 1

Pulse Generator

Output pulses:

if (t >= PhaseDelay) && Pulse is on
  Y(t) = Amplitude
else
  Y(t) = 0
end

Pulse type determines the computational technique used.

Time-based is recommended for use with a variable step solver, while Sample-based is recommended for use with a fixed step solver or within a discrete portion of a model using a variable step solver.

Parameters

Pulse type: Time based

Time (t):   Use simulation time

Amplitude:

4

Period (secs):

0.02

Pulse Width (% of period):

50

Phase delay (secs):

0*0.02/360

OK   Cancel   Help   Apply

Figure 7.41: Pulse 1 used for first IGBT.

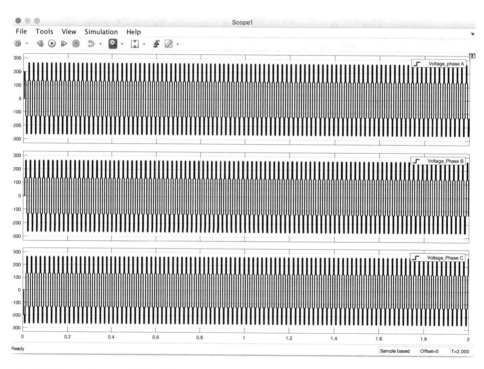

Figure 7.42: Three phases of the alternating voltage output.

Bibliography

[1] Alciatore, D. and Histand, H. B. *Mechatronics*, McGraw-Hill, 2005. 4

[2] Ashley, S. Getting a hold on mechatronics, *Mechanical Engineering, ASME Magazine*, 1997. 1

[3] Brown, F. T. *Engineering System Dynamics: A Unified Graph Centered Approach*, Marcel Dekker, 2001. DOI: 10.1201/b18080. 4

[4] Bolton, W. *Mechatronics: Electronic Control Systems in Mechanical and Electrical Engineering*, Longman Publishers, 2015. 4

[5] Cetinkunt, S. *Mechatronics*, John Wiley, 2007. 4

[6] Das, S. *Mechatronic Modeling and Simulation Using Bond Graph*, CRC Press, 2009. DOI: 10.1201/b15831. 4

[7] Definitions of Mechatronics, http://mechatronics.colostate.edu/definitions/ 1

[8] De Silva, C. W. *Mechatronics, an Integrated Approach*, CRC Press, 2005. DOI: 10.1201/b12787. 4

[9] De Silva, C. W. *Mechatronics, a Foundation Course*, CRC Press, 2010. DOI: 10.1201/9781420082128.

[10] Karno, D. C., Margolis, D. L., and Rosenberg, R. C. *System Dynamics: Modeling and Simulation of Mechatronic Systems*, John Wiley & Sons, 2012. DOI: 10.1002/9781118152812. 4, 5

[11] Shetty, D. and Kolk, R. *Mechatronics System Design*, PWS Publishing, 2010. 4

Author's Biography

SHUVRA DAS

Shuvra Das is a Professor of Mechanical Engineering at University of Detroit Mercy and Director of International Programs in the college of Engineering and Science. He has an undergraduate degree in Mechanical Engineering from Indian Institute of Technology, and a Master's and Ph.D. in Engineering Mechanics from Iowa State University. He was a post-doctoral researcher at University of Notre Dame and also worked in industry prior to being in academia for 25 five years. He has taught a variety of courses ranging from freshmen to advanced graduate level within Mechanical Engineering, as well as in the Advanced Electric Vehicle program. His research interests include mechatronics system modeling and simulation, multi-physics process and product simulation using CAE tools such as Finite and Boundary Elements, structural analysis and design, design processes and design thinking, and engineering education. He received many awards for teaching and research at Detroit Mercy as well as from organizations outside the university. He served as Associate Dean for Research and Outreach at Detroit Mercy for six years. During this time, he launched the undergraduate Robotics and Mechatronics Systems Engineering Program, the Advanced Electric Vehicle certificate program, and several additional engineering degree programs in collaboration with Chinese Universities.

Printed in the United States
by Baker & Taylor Publisher Services